# THE O[illegible] CONTINENT[illegible] OCEANS

BY

ALFRED WEGENER

*Translated from the Fourth Revised German Edition by*

*JOHN BIRAM*

DOVER PUBLICATIONS, INC.
NEW YORK

Published in Canada by General Publishing Company, Ltd., 30 Lesmill Road, Don Mills, Toronto, Ontario.

Published in the United Kingdom by Constable and Company, Ltd., 10 Orange Street, London WC 2.

This Dover edition, first published in 1966, is a new English translation of the 1962 printing of the fourth revised edition of *Die Entstehung der Kontinente und Ozeane* published in 1929 by Friedr. Vieweg & Sohn. It is published by special arrangement with Friedr. Vieweg & Sohn, Braunschweig.



*Library of Congress Catalog Card Number: 66-28270*

Manufactured in the United States of America
Dover Publications, Inc.
180 Varick Street
New York, N.Y. 10014

# *Alfred Wegener*

ALFRED WEGENER was born in Berlin on the 1st of November 1880, the youngest child of the evangelical preacher Dr. Richard Wegener and his wife Anna, née Schwarz. He attended the Köllnisches Gymnasium in Berlin and later studied at the Universities of Heidelberg, Innsbruck and Berlin. At the completion of his studies he entered the "Urania" at Berlin as an astronomer. However, he soon became a second technical assistant to his brother Kurt at the Prussian Aeronautical Observatory in Tegel. The two brothers carried out a joint balloon flight of $52\frac{1}{2}$ hours, a record for that time; the flight began in Berlin and continued over Jutland and the Kattegat and then towards the Spessart area of Germany. The journey provided a test of the accuracy of the spirit-level clinometer as an instrument for flight navigation.

In 1906, Alfred Wegener went with a Danish national expedition to the northeast coast of Greenland for two years; on this expedition he learned the technique of polar travel. His published observations related essentially to meteorological questions. After returning from Greenland he became a lecturer in astronomy and meteorology at the University of Marburg. His lectures were the foundations of his textbook *Thermodynamik der Atmosphäre,* which went to three editions but is now out of print. In accordance with Alfred Wegener's plans, it was replaced by the book *Vorlesungen über Physik der Atmosphäre,* by Alfred and Kurt Wegener, published in 1935.

In 1912, together with J. P. Koch, Wegener undertook his second expedition to Greenland, with the purpose of spending a winter at the eastern edge of the inland ice, and then crossing Greenland at its widest part. The expedition was almost completely wrecked during the ascent of the inland glacier by intensive calving of the ice, which extended up to the encampment area. The crossing took place in 1913, after wintering, and lasted two months. The expedition was only just able to reach the west coast.

In 1914 Wegener was drafted as a reserve lieutenant of the Queen Elisabeth Grenadier Guards' Third Regiment and assigned to the field regiment. During the advance into Belgium he was shot through the arm. About fourteen days after his return to duty, a bullet lodged in his neck. As a result of this he was never fit for active duty again and was only employed in the field meteorological service. In 1915 the first edition of his book *Die Entstehung der Kontinente und Ozeane* appeared. This book was concerned with re-establishing the connection between geophysics on the one hand and geography and geology on the other, a connection which had been completely broken by specialist development of these branches of science. The second edition appeared in 1920, the third in 1922 and the fourth in 1929. Each edition was a complete revision, involving material compiled as a result of criticism, initially adverse, but subsequently interested. The third edition was translated into French in 1924 by M. Reichel under the title *La genèse des continents et des océans* and was published as a volume in the Librarie Scientifique Albert Blanchard, Paris. This edition was also translated into English by J. G. A. Skerl in the same year (*The Origin of Continents and Oceans*) with a foreword by the President of the English Geological Society, John W. Evans, C.B.E., F.R.S. This translation was published by Methuen & Co. Ltd., London. A Spanish translation of the third edition also came out in the same year, entitled *La génesis de los continentes y océanos*. The translator was Vicente Inglada Ors; the publisher, Biblioteca de la Revista de Occidente, Madrid. In 1925 G. F. Mirtzinka (Moscow and Leningrad) published a translation by Marii Mirtzink. In 1924 the work was supplemented by *Die Klimate der geologischen Vorzeit* by W. Köppen and A. Wegener (Verlag Gebrüder Bornträger).

After the war, Alfred, like his brother Kurt, became a departmental head at the German Marine Observatory in Hamburg, and he was also a lecturer in meteorology at the newly founded University of Hamburg. In 1924 he accepted an appointment as Professor of Meteorology and Geophysics at Graz University (Austria).

Wegener had planned a new Greenland expedition in collaboration with J. P. Koch, for 1928. The latter died in 1928 and this meant that the expedition had to be planned as a purely German affair. Wegener received the strong support of the German Research Association (Notgemeinschaft der Deutschen Wissenschaft; His Excellency Herr Schmidt-Ott, President). In 1929 he first of all clarified the question of the most favourable route up the inland icecap from the west coast. The main expedition began in 1930. Perhaps the most important result of the expedition was the discovery that the thickness of inland ice is more than 1800 metres.

In November 1930, Alfred Wegener met his death on the inland icecap.

Wegener had already decided by 1928 that a new revision of his book would be beyond him because the literature relevant to the problem had become too extensive and specialised for a single worker to survey. It was therefore his wish that any further edition that might prove necessary should appear without alteration.

Kurt Wegener

## *Publisher's Note to the Last German Edition*

In accordance with the author's express wish, and in appreciation of the great historical significance of this scientific document, we present the unrevised text of the fourth edition, as we did in the case of the fifth and sixth. We have avoided supplementation of the reference list, as was undertaken for the last editions, especially since the literature has increased meanwhile to an enormous extent. However, we did not want to omit the account of the life and work of Alfred Wegener prepared by his brother Kurt for the fifth edition. In all other respects this book is offered once again exactly as it came from the author's own hand.

*Winter 1961* Editor and Publisher

## *Foreword*

SCIENTISTS still do not appear to understand sufficiently that all earth sciences must contribute evidence towards unveiling the state of our planet in earlier times, and that the truth of the matter can only be reached by combining all this evidence.

The well-known South African geologist du Toit wrote quite recently [78]: "As already stated, we must turn almost exclusively to the *geological* evidence to decide the probability of this hypothesis (continental drift), because arguments based on such matters as the distribution of fauna are not competent here; they can generally be explained equally well, even if less neatly, by the orthodox view that assumes the existence of extended land bridges, later sunk below sea level."

On the other hand, the palæontologist von Ihering [122] is short and to the point: "It is not my job to worry about geophysical processes." He holds to the "conviction that only the *history of life* on the earth enables one to grasp the geographical transformations of the past."

I myself in a weak moment once wrote of the drift theory [121]: "For all that, I believe that the final resolution of the problem can only come from *geophysics*, since only that branch of science provides sufficiently precise methods. Were geophysics to come to the conclusion that the drift theory is wrong, the theory would have to be abandoned by the systematic earth science as well, in spite of all corroboration, and another explanation for the facts would have to be sought."

It would be easy to add to the list of such opinions, each scientist deeming his own field to be the one most competent, or indeed the only one competent, to judge the issue.

In fact, however, the situation is obviously quite otherwise. At a specified time the earth can have had just one configuration. But the earth supplies no direct information about this. We are like a judge confronted by a defendant who declines to answer, and we must determine the truth from the circumstantial evidence. All the proofs we can muster have the deceptive character of this type of evidence. How would we assess a judge who based his decision on part of the available data only?

It is only by combining the information furnished by all the earth sciences that we can hope to determine "truth" here, that is to say, to find the picture that sets out all the known facts in the best arrangement and that therefore has the highest degree of probability. Further, we have to be prepared always for the possibility that each new discovery, no matter which science furnishes it, may modify the conclusions we draw.

This conviction gave me the stimulus to continue at times when my spirits failed me during the revision of this book. For it is beyond one man's power to follow up completely the details of the snowballing literature on drift theory in the various sciences. In spite of all my efforts, many gaps, even important ones, will be found in this book. That I was able to achieve the degree of comprehensiveness I did is due solely to the very large number of communications received from scientists in all the relevant fields, and I am most grateful for them.

The book is addressed equally to geodesists, geophysicists, geologists, palæontologists, zoogeographers, phytogeographers and palæoclimatologists. Its purpose is not only to provide research workers in these fields with an outline of the significance and usefulness of the drift theory as it applies to their own areas, but also mainly to orient them with regard to the applications and corroborations which the theory has found in areas other than their own.

Everything of interest concerning the history of this book, which is also the history of the drift theory, will be found in the first chapter.

The reader is referred to the Appendix for evidence of a shift of North America brought out by the new determinations of longitude in 1927; this result first appeared during the time the book was in proof.

*Graz, November 1928* ALFRED WEGENER

*Contents*

Alfred Wegener

CHAPTER 1

# *Historical Introduction*

THE BACKGROUND to this book may not be wholly without interest. The first concept of continental drift first came to me as far back as 1910, when considering the map of the world, under the direct impression produced by the congruence of the coastlines on either side of the Atlantic. At first I did not pay attention to the idea because I regarded it as improbable. In the fall of 1911, I came quite accidentally upon a synoptic report in which I learned for the first time of palæontological evidence for a former land bridge between Brazil and Africa. As a result I undertook a cursory examination of relevant research in the fields of geology and palæontology, and this provided immediately such weighty corroboration that a conviction of the fundamental soundness of the idea took root in my mind. On the 6th of January 1912 I put forward the idea for the first time in an address to the Geological Association in Frankfurt am Main, entitled "The Geophysical Basis of the Evolution of the Large-scale Features of the Earth's Crust (Continents and Oceans)" ("Die Herausbildung der Grossformen der Erdrinde (Kontinente und Ozeane) auf geophysikalischer Grundlage"). A second address followed, this one on the 10th of January, delivered before the Society for the Advancement of Natural Science in Marburg under the title "Horizontal Displacements of the Continents" ("Horizontalverschiebungen der Kontinente"). In the same year, the two first publications also appeared [1, 2]. Further work on the theory was prevented by my participation in the crossing of Greenland led by J. P. Koch in 1912/1913, and later by war service. However, in 1915 I was able to make use of a prolonged sick-leave to furnish a rather more detailed account, with the same title as this volume and published by Vieweg [3]. When, after the end of the war, a second edition (1920) became necessary, the publisher was kind enough to transfer the book from the Sammlung Vieweg to the Sammlung Wissenschaft (Science Series); this made a more thoroughgoing revision possible. In 1922 appeared the third edition, again fundamentally improved, and in an unusually large printing so that I could work on other problems for a few years. It has been completely out of print for some time. A series of translations of this edition appeared, two Russian, one English, one French, one Spanish and one Swedish. I undertook to make a few changes in the German text for the Swedish translation, which appeared in 1926.

This fourth edition of the German original has once again been thoroughly revised; in fact, it has taken on an almost totally different character from its predecessors. When the previous edition was being written, there was already a comprehensive literature on continental drift which had to be taken into account. However, this literature was confined in the main to expressions of agreement or disagreement and to the citing of individual observations which spoke out or appeared to speak out either for or against the correctness of the theory; whereas since 1922, not only has the discussion of this question within the different earth sciences grown out of all proportion, but the very character of the discussion has altered to some extent. The theory is being used more and more as a basis for more extensive investigations. In addition, there is the recent precise evidence for the present-day shift of Greenland, which for many people has probably placed the discussion on a completely new footing. Therefore, while the earlier editions contained in essence merely a presentation of the theory itself and a collection of the individual facts in support of it, the present edition represents a transitional stage between the mere presentation of the theory and a synoptic exposition of these new branches of research.

Even when I was first occupied with this question, and also from time to time during the later development of the work, I encountered many points of agreement between my own views and those of earlier authors. As far back as 1857 Green spoke of "segments of the earth's crust which float on the liquid core" [63]. Rotation of the whole crust—whose components were supposed not to alter their relative positions—has already been assumed by several writers, such as Löffelholz von Colberg [4], Kreichgauer [5], Evans and others.

H. Wettstein wrote a book [6] in which (besides many inanities) the idea of large horizontal relative displacements of the continents is to be found. In his view, the continents—whose shelves he did not take into account—undergo not only displacement, but also deformation; they all drift westwards under tidal forces of the sun acting on the viscous material of the earth (an idea also held by E. H. L. Schwarz [7]). However, Wettstein, too, regarded the oceans as sunken continents, and he expressed fantastic views, which we pass over here, on the so-called geographical homologies and other problems of the earth's surface. Like myself, Pickering started out from the congruence of the southern Atlantic coastlines in a work [8] in which he expressed the supposition that America had broken away from Europe-Africa and was dragged the breadth of the Atlantic. However, he did not observe that one must in fact assume that an earlier connection between the two continents existed during their geological history up to the Cretaceous period, and he therefore assigned this connection to a dim and distant past, believing the breakaway to be bound up with G. H. Darwin's assumption that the moon was flung from the earth, and that traces of this can still be seen in the Pacific basin.

In a short article in 1909 Mantovani [86] expressed some ideas on continental displacement and explained them by means of maps which differ in part from mine but at some points agree astonishingly closely: for example, in regard to the earlier grouping of the southern continents around southern Africa. It was pointed out to me in correspondence that Coxworthy, in a book which appeared after 1890, put forward the hypothesis that today's continents are the disrupted parts of a once-coherent mass [9]. I have had no opportunity to examine the book.

I also discovered ideas very similar to my own in a work of F. B. Taylor's [10] which appeared in 1910. Here, he assumed by no means inconsiderable horizontal shifts of the individual continents in Tertiary times, and connected these with the large Tertiary systems of folding. He came to virtually the same conclusions as my own, for example, about the separation of Greenland from North America. In the case of the Atlantic, he assumed that only part of its width is due to drag displacement of the American land mass and that the rest is due to submergence and constitutes the mid-Atlantic ridge. This viewpoint, too, differs only quantitatively from my own, but not in crucial or novel ways. For this reason, Americans have sometimes called the drift theory the Taylor-Wegener theory. However, I have received the impression when reading Taylor that his main object was to find a formative principle for the arrangement of the large mountain chains and believed this to be found in the drift of land from polar regions; my impression is therefore that in Taylor's train of thought continental drift in our sense played only a subsidiary role and was given only a very cursory explanation.

I myself only became acquainted with these works—including Taylor's—at a time when I had already worked out the main framework of drift theory, and some of them I encountered much later on. It is of course not beyond the bounds of possibility that further works will be discovered in the course of time which will prove to contain elements of agreement with drift theory or to have anticipated a point here or there. Historical investigations have not been undertaken as yet and are not intended in the present book.

CHAPTER 2

# *The Nature of the Drift Theory and Its Relationship to Hitherto Prevalent Accounts of Changes in the Earth's Surface Configuration in Geological Times*

It is a strange fact, characteristic of the incomplete state of our present knowledge, that totally opposing conclusions are drawn about prehistoric conditions on our planet, depending on whether the problem is approached from the biological or the geophysical viewpoint.

Palæontologists as well as zoo- and phytogeographers have come again and again to the conclusion that the majority of those continents which are now separated by broad stretches of ocean must have had land bridges in prehistoric times and that across these bridges undisturbed interchange of terrestrial fauna and flora took place. The palæontologist deduces this from the occurrence of numerous identical species that are known to have lived in many different places, while it appears inconceivable that they should have originated simultaneously but independently in these areas. Furthermore, in cases where only a limited percentage of identities is found in contemporary fossil fauna or flora, this is readily explained, of course, by the fact that only a fraction of the organisms living at that period is preserved in fossil form and has been discovered so far. For even if the whole groups of organisms on two such continents had once been absolutely identical, the incomplete state of our knowledge would necessarily mean that only part of the finds in both areas would be identical and the other, generally larger, part would seem to display differences. In addition, it is obviously the case that even where the possibility of interchange was unrestricted, the organisms would not have been quite identical in both continents; even today Europe and Asia, for example, do not have identical flora and fauna by any means.

Comparative study of *present-day* animal and plant kingdoms lead to the same result. The species found today on two such continents are indeed different, but the genera and families are still the same; and what is today a genus or family was once a species in prehistoric times. In this way the relationships between present-day terrestrial faunas and floras lead to the conclusion that they were once identical and that therefore there must have been exchanges, which could only have taken place over a wide land bridge. Only after the land bridge had been broken were the floras and faunas subdivided into today's various species. It is probably not an exaggeration to say that if we do not accept the idea of such former land connections, the whole evolution of life on earth and the affinities of present-day organisms occurring even on widely separated continents must remain an insoluble riddle.

Here is just one testimony amongst many: de Beaufort wrote [123]: "Many other examples could be given to show that it is impossible in zoogeography to arrive at an acceptable explanation of the distribution of animals if no connections between today's separate continents are assumed to have existed, and not only land bridges from which, as Matthew put it, only a few planks have been removed, but also such that joined land masses now separated by deep oceans."[1]

Obviously, there are many individual questions which are insufficiently explained by this theory. In many cases former land bridges have been assumed on the basis of very meagre evidence and have not been confirmed by the advance of research. In other cases there is still no complete agreement on the point in time when the connection was broken and the present-day separation began. However, in the case of the most important of these ancient land bridges, there does

[1] Arldt [135] states: "Of course, there are today still some opponents of the land-bridge theory. Among them, G. Pfeffer is worth special mention. He starts from the point that various forms now restricted to the southern hemisphere are manifest as fossils in the northern hemisphere. This precludes any doubt, he says, that these forms were once more or less universally distributed. If this conclusion is not completely compelling, still less is the further conclusion that we should assume a universal distribution even in all cases where there is a discontinuous distribution in the south but no fossil evidence as yet in the north. If he wants to explain distribution anomalies solely by migrations between the northern continents and their mediterranean bridges, the assumption rests on a very uncertain footing." That the affinities found on the southern continents can be explained more *simply* and *completely* by direct land bridges than by parallel migrations from the common northern region will require no further comment, even though in individual cases the processs could have been the one that Pfeffer assumed.

already exist today a gratifying unanimity among specialists, whether they base their conclusions on geographical distribution of the mammals or earthworms, on plants or on some other portion of the world of organisms. Arldt [11], using the statements or maps of twenty scientists,[2] has drawn up a sort of table of votes for or against the existence of the different land bridges in the various geological periods. For the four chief bridges, I have presented the results graphically in Figure 1. Three curves are shown for each

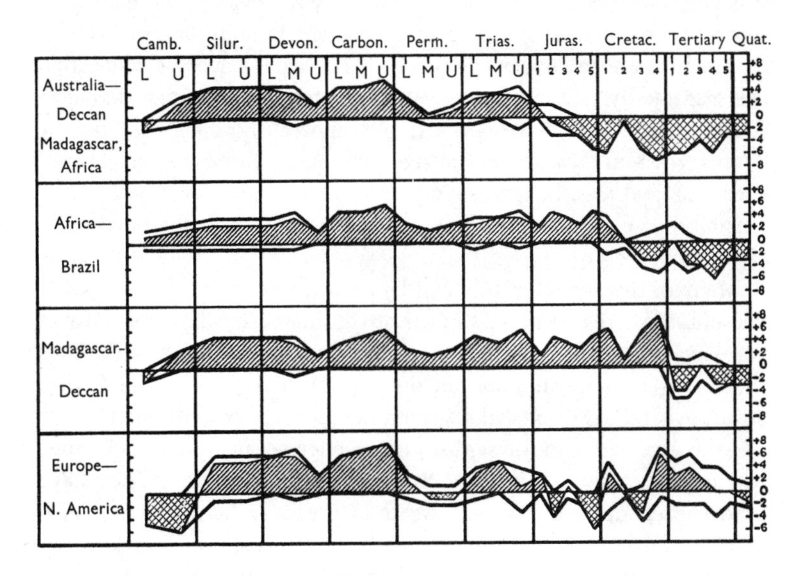


FIG. 1. The number of proponents (upper curves) and opponents (lower curves) of the existence of four land bridges since Cambrian times.

The difference (majority) is hatched, and crosshatched when the majority opposes.

bridge—the number of yeas, the number of nays and the difference between them, i.e., the strength of the majority vote, which is emphasised by hatching the appropriate area. Thus, the top section indicates that according to the majority of researchers the bridge between Australia on the one side and India, Madagascar and Africa

[2] Arldt, Burckhardt, Diener, Frech, Fritz, Handlirsch, Haug, von Ihering, Karpinsky, Koken, Kossmat, Katzer, Lapparent, Matthew, Neumayr, Ortmann, Osborn, Schuchert, Uhlig and Willis.

(ancient "Gondwanaland") on the other lasted from Cambrian times to the beginning of the Jurassic, but was then disrupted. The second section shows that the old bridge between South America and Africa ("Arch-helenis") is considered by most to have broken in the Lower to Middle Cretaceous. Still later, at the transition between Cretaceous and Tertiary, the old bridge between Madagascar and the Deccan ("Lemuria") is assumed by the majority to have broken (see section 3 of Fig. 1). The land bridge between North America and Europe was very much more irregular, as shown by section 4. But even here there is a substantial measure of agreement in spite of the frequent change in the behaviour of the curves. In earlier times the connection was repeatedly disturbed, i.e., in the Cambrian, Permian and also Jurassic and Cretaceous periods, but apparently only by shallow "transgressions," which permitted subsequent re-formation. However, the final breach, corresponding now to a broad stretch of ocean, can only have occurred in the Quaternary, at least in the north near Greenland.

Many of the details of this will be treated later in the book. Only one point is stressed here, so far not considered by the exponents of the land-bridge theory, but of great importance: These former land bridges are postulated not only for such regions as the Bering Strait, where today a shallow continental-shelf sea, or floodwater fills the gap, but also for regions now under ocean waters. All four examples in Figure 1 involve cases of this latter type. They have been chosen deliberately because it is precisely here that the new concept of drift theory begins, as we have yet to show.

Since it was previously taken for granted that the continental blocks—whether above sea level or inundated—have retained their mutual positions unchanged throughout the history of the planet, one could only have assumed that the postulated land bridges existed in the form of intermediate continents, that they sank below sea level at the time when interchange of terrestrial flora and fauna ceased and that they form the present-day ocean floors between the continents. The well-known palæontological reconstructions arose on the basis of such assumptions, one example of them, for the Carboniferous, is given in Figure 2.

This assumption of sunken intermediate continents was in fact the most obvious so long as one based one's stand on the theory of the contraction or shrinkage of the earth, a viewpoint we shall have to examine more closely in what follows. The theory first appeared in

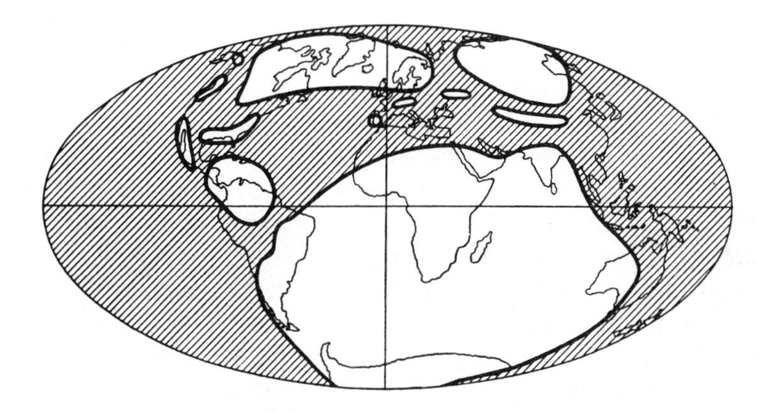

FIG. 2. Distribution of water (hatched) and land in the Carboniferous, according to the usual conception.

Europe. It was initiated and developed by Dana, Albert Heim and Eduard Suess in particular, and even today dominates the fundamental ideas presented in most European textbooks of geology. The essence of the theory was expressed most succinctly by Suess: "The collapse of the world is what we are witnessing" [12, Vol. 1, p. 778]. Just as a drying apple acquires surface wrinkles by loss of internal water, the earth is supposed to form mountains by surface folding as it cools and therefore shrinks internally. Because of this crustal contraction, an overall "arching pressure" is presumed to act over the crust so that individual portions remain uplifted as horsts. These horsts are, so to speak, supported by the arching pressure. In the further course of time, these portions that have remained behind may sink faster than the others and what was dry land can become sea floor and vice-versa, the cycle being repeated as often as required. This idea, put forth by Lyell, is based on the fact that one finds deposits from former seas almost everywhere on the continents. There is no denying that this theory provided historic service in furnishing an adequate synthesis of our geological knowledge over a long period of time. Furthermore, because the period was so long, contraction theory was applied to a large number of individual research results with such consistency that even today it possesses a degree of attractiveness, with its bold simplicity of concept and wide diversity of application.

Ever since our geological knowledge was made the subject of that impressive synthesis, the four volumes by Eduard Suess entitled *Das Antlitz der Erde*, written from the standpoint of contraction theory, there has been increasing doubt as to the correctness of the basic idea. The conception that all uplifts are only apparent and consist merely of remnants left from the general tendency of the crust to move towards the centre of the earth, was refuted by the detection of absolute uplifts [71]. The concept of a continuous and ubiquitous arching pressure, already disputed on theoretical grounds for the uppermost crust by Hergesell [124] has proved to be untenable because the structure of eastern Asia and the eastern African rift valleys have, on the contrary, enabled one to deduce the existence of tensile forces over large portions of the earth's crust. The concept of mountain folding as crustal wrinkling due to internal shrinkage of the earth led to the unacceptable result that pressure would have to be transmitted inside the earth's crust over a span of 180 great-circle degrees. Many authors, such as Ampferer [13], Reyer [14], Rudzki [15] and Andrée [16], among others, have opposed this quite rightly, claiming that the surface of the earth would have to undergo regular overall wrinkling, just as the drying apple does. However, it was particularly the discovery of the scale-like "sheet-fault structure" or overthrusts in the Alps which made the shrinkage theory of mountain formation, which presented enough difficulties in any case, seem more and more inadequate. This new concept of the structure of the Alps and that of many other ranges, which was introduced by the works of Bertrand, Schardt, Lugeon and others, leads to the idea of far larger compressions than did the earlier theory. Following previous ideas, Heim calculated in the case of the Alps a 50% contraction, but on the basis of the sheet-faulting theory, now generally accepted, contraction of $\frac{1}{4}$ to $\frac{1}{8}$ of the initial span [17]. Since the present-day width of the chain is about 150 km, a stretch of crust from 600 to 1200 km wide (5–10 degrees of latitude) must have been compressed in this case. Yet in the most recent large-scale synthesis on Alpine sheet-faults, R. Staub [18] agrees with Argand that the compression must have been even greater. On page 257 he concludes:

"The Alpine orogenesis is the result of the northward drift of the African land mass. If we smooth out only the Alpine folds and sheets over the transverse section between the Black Forest and Africa, then in relation to the present-day distances of about 1800 km, the original distance separating the two must have been about 3000 to 3500 km, which means an alpine compression (in the wider sense of the word Alpine) of around 1500 km. Africa must have been displaced relative to Europe by this amount. What is involved here is a true continental drift of the African land mass and an extensive one at that."[3]

Other geologists have put forward similar views, as for example F. Hermann [106], E. Hennig [19] or Kossmat [21], who states "that the formation of mountains must be explained by large-scale tangential movements of the crust, which cannot be incorporated in the scope of the simple contraction theory." In the case of Asia, Argand [20], especially, has developed an analogous theory in the course of a comprehensive investigation to which we shall return later. He and Staub have done the same for the case of the Alps. No attempt to relate these enormous compressions of the crust to a drop in temperature of the earth's core can be anything but a failure.

Moreover, even the apparently obvious basic assumption of contraction theory, namely that the earth is continuously cooling, is in full retreat before the discovery of radium. This element, whose decay produces heat continuously, is contained in measurable amounts everywhere in the earth's rock crust accessible to us. Many measurements lead to the conclusion that even if the inner portion had the same radium content, the production of heat would have to be incomparably greater than its conduction outwards from the centre, which we can measure by means of the rise of temperature with depth in mines, taking into account the thermal conductivity of rock. This would mean, however, that the temperature of the earth must rise continuously. Of course, the very low radioactivity of iron meteorites suggests that the iron core of the earth presumably contains much less radium than the crust, so that this paradoxical conclusion can

[3] It seems that estimations of the size of the Alpine compression are always on the increase. Staub wrote recently [214, similarly in 215]: "If we now, however, imagine these Alpine sheets, which are probably stacked twelvefold, to be smoothed out again . . ., the solid Alpine hinterland would necessarily lie much further south, and the original distance between foreland and hinterland would probably have been ten to twelve times greater than it is today." He adds: "Formation of a mountain range therefore originates quite clearly and certainly from independent drifting of larger blocks, surely continental blocks by their structure and composition; and thus, starting from Alpine geology and Hans Schardt's sheet theory, we arrive quite obviously and naturally at the acknowledgment of the basic principle of the great Wegener theory of continental drift."

perhaps be avoided. In any case, it is no longer possible, as it once was, to consider the thermal state of the earth as a temporary phase in the cooling process of a ball that was formerly at a higher temperature. It should be regarded as a state of equilibrium between radioactive heat production in the core and thermal loss into space. In fact, the most recent investigations into this question, which will be discussed in more detail later on, imply that actually, at least under the continental blocks, more heat is generated than is conducted away, so that here the temperature must be rising, though in the ocean basins conduction exceeds production. These two processes lead to equilibrium between production and loss rate, taking the earth as a whole. In any case, one can see that through these new views the foundation of the contraction theory has been completely removed.

There are still many other difficulties which tell against the contraction theory and its mode of thinking. The concept of an unlimited periodic interchange between continent and sea floor, which was suggested by marine sediments on present-day continents, had to be strictly curtailed. This is because more precise investigation of these sediments showed with increasing clarity that what was involved was coastal-water sediments, almost without exception. Many sedimentary deposits formerly claimed as oceanic proved to be coastal; one example is chalk, as proved by Cayaux. Dacqué [22] has given a good review of the problem. Only in the case of a very few types of sediment, such as the low-lime Alpine radiolarites and certain red clays reminiscent of the red deep-sea clay, is formation in deep waters (4–5 km) still assumed today, particularly since sea water dissolves out lime only at great depths. However, the area of these true deep-sea deposits on present-day continents is so tiny compared with the areas of the continents and the areas of coastal water sediments on them that the theory of the basically shallow-water nature of marine fossil deposits on present-day continents is unaffected. For the contraction theory, however, a considerable difficulty arises. Since coastal shallows must be counted, geophysically, as part of the continental blocks, the nature of these marine fossils implies that these blocks have been "permanent" throughout the history of the earth and have never formed ocean floors. Are we then still to assume that today's sea floors were ever continents? The justification for this conclusion is obviously removed by establishing that the marine sediments found on continents were formed in shallows. But more than this, the conclusion now leads to an open contradiction. If we reconstruct intercontinental bridges of the type shown in Figure 2, thus filling up a large part of today's ocean basins without having the possibility of compensating for this by submergence of present-day continental regions to the sea-floor level, there would be no room for the volume of the world's oceans in the now much reduced deep-sea basins. The water displacement of the intercontinental bridges would be so enormous that the level of the world's oceans would rise above that of the whole continental area of the earth and all would be flooded, today's continents and the bridges alike. The reconstruction would not therefore achieve the desired end, i.e., dry land bridges between continents. Figure 2 therefore represents an impossible reconstruction unless we introduce further hypotheses which are "ad hoc" improbabilities; for example, that the mass of ocean water was exactly the required amount less at the former period than it is today, or that the deep-sea basins remaining at that time were precisely the required amount deeper than today. Willis and A. Penck, among others, have brought up this peculiar difficulty.

Of the many objections to contraction theory, one more only will be emphasised; it has very special importance. Geophysicists have decided, mainly on the basis of gravity determinations, that the earth's crust floats in hydrostatic equilibrium on a rather denser, viscous substrate. This state is known as *isostasy*, which is nothing more than hydrostatic equilibrium according to Archimedes' principle, whereby the weight of the immersed body is equal to that of the fluid displaced. The introduction of a special word for this state of the earth's crust has some point because the liquid in which the crust is immersed apparently has a very high viscosity, one which is hard to imagine, so that oscillations in the state of equilibrium are excluded and the tendency to restore equilibrium after a perturbation is one which can only proceed with extreme slowness, requiring many millennia to reach completion. Under laboratory conditions, this "liquid" would perhaps scarcely be distinguishable from a "solid." However, it should be remembered here that even with steel, which we certainly consider a solid, typical flow phenomena occur, just before rupture, for example.

An example of perturbation of isostasy of the crust is shown by the load to which an inland icecap subjects it. The result is that the crust slowly sinks under this load and tends towards a new equilibrium position to correspond with the loading. When the icecap has melted, the original position of equilibrium is gradually resumed, and the shore lines formed during the process of depression are elevated along with the crust. The "isobase charts" of de Geer [23], drawn up from the shore lines, show for the last glaciation of Scandinavia a central depression of at least 250 m, gradually decreasing towards the perimeter; for the most extensive of the Quaternary glaciations still higher values must be assumed. In Figure 3 we reproduce a chart

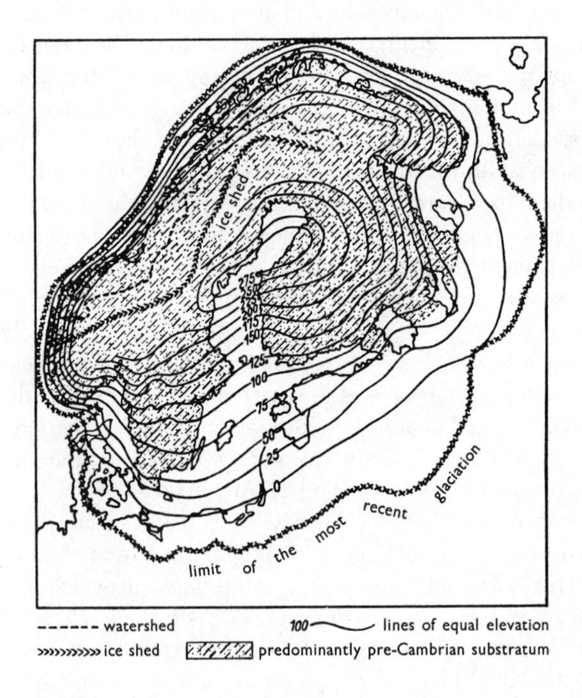


FIG. 3. Post-glacial elevation contours (in metres) for Fennoscandia (according to Högbom).

of this post-glacial elevation of "Fennoscandia" (Finland, Sweden and Norway) according to Högbom (taken from Born [43]). The same phenomenon has been proved by de Geer to have occurred for the glaciated region of North America. Rudzki [15] has shown that, assuming isostasy, plausible values for the thickness of inland ice layers can be calculated, i.e., 930 m for Scandinavia and 1670 m for North America, where the depression amounted to 500 m. Because of the viscosity of the substrate the equilibration movements naturally lag far behind: the shore lines generally formed only after the melting of the ice, but before the elevation of the land, and even today Scandinavia is still rising by about 1 m in 100 years, as shown by tide-gauge readings.

Even sedimentary deposits result in a subsidence of the blocks, as Osmond Fisher was probably the first to recognise: every deposition from above leads to a subsidence of the block, somewhat delayed of course, so that the new surface occupies almost the same level as the old. In this way many kilometres' thickness of deposit can arise and yet all the layers are formed in shallow water.

Later on we shall examine the theory of isostasy more closely. Here we shall simply say that it has been established by geophysical observations over so wide a range that it is now part of the solid foundation of geophysics and its basic truth can no longer be doubted.[4]

One can see immediately that this result runs quite counter to the ideas of contraction theory and that it is very hard to combine one with the other. In particular, it seems impossible, in view of the isostatic principle, that a continental block the size of a land bridge of required size could sink to the ocean bottom without a load or that the reverse should happen. Isostasy is therefore in contradiction not only to contraction theory, but in particular also to the theory of sunken land bridges as derived from the distribution of organisms.[5]

[4] Americans, e.g., Taylor [101], sometimes mean by "isostasy" Bowie's theory of the origin of geosynclines and mountain ranges. According to Bowie [224], the initial elevation of sediment-filled basins, the geosynclines, comes from a rise in their isotherms, and hence a volumetric expansion. Once this has led to a land elevation, erosion sets in and a jagged mountain range is formed, whose substrate continually rises due to reduction in loading. Finally, the isotherms are raised to an abnormal height by this elevation, and begin to move slowly downwards; the block cools and contracts and the surface sinks; a depression is formed from the mountain region and renewed sedimentation occurs. This produces further depression or subsidence until the isotherms are abnormally low in level, then rise again, and so on over many cycles. This concept, which cannot be applied to the great folded ranges with their overthrusts, as Taylor and others have emphasised, does indeed make use of the principle of isostasy but should not be given the simple title of "the theory of isostasy."

[5] The objections to the contraction theory enumerated here are mainly directed against its typical earlier form. Very recently, attempts have been made to modernise the theory and to answer the objections, partly by restricting it and partly by adding hypotheses; various authors have been involved, such as Kober [24], Stille [25], Nölcke [26], and Jeffreys [102], among others. This is also true of the theory publicised by R. T. Chamberlin [160] which supposes contraction to be caused by "rearrangement" of material in the earth resulting from the planetesimal origin of the earth accepted by this author. Although one cannot deny

In the foregoing, we deliberately reviewed the objections to contraction theory in some detail. This is because in one part of the train of thought discussed here another theory is rooted; this is known as the "theory of permanence" and is especially widespread among American geologists. Willis [27] formulated it as follows: "The great ocean basins constitute permanent features of the earth's surface, and have with little change in shape occupied the same positions as now since the ocean waters were first gathered." In fact, we have already referred above to the fact that the marine sediments on present-day continents were formed in shallow waters, and we deduced that the continental blocks as such have been permanent throughout the earth's history. Isostasy theory proves the impossibility of regarding present-day ocean floors as sunken continents, and this extends the scope of the result based on marine sediments to comprise a general permanence of deep-sea floors and continental blocks. Further, since here, too, the apparently obvious assumption was made that the continents have not changed their relative positions, Willis's formulation of the "permanence theory" appears to be a logical conclusion from our geophysical knowledge, disregarding, of course, the postulate of former land bridges, derived from the distribution of organisms. So we have the strange spectacle of two quite contradictory theories of the prehistoric configuration of the earth being held simultaneously—in Europe an almost universal adherence to the idea of former land bridges, in America to the theory of the permanence of ocean basins and continental blocks.

It is probably no accident that the permanence theory has its most numerous adherents in America: geology developed late there—thus simultaneously with geophysics—and this necessarily led to more rapid and complete adoption by geology of the results advanced by its sister science than in Europe. There was absolutely no temptation to make the contraction theory, which contradicts geophysics, one of the basic assumptions. It was quite otherwise in Europe, where geology already had a long period of development behind it before geophysics had produced its first results, and had, without benefit of geophysics, already arrived at an overall view of

---

that these attempts show a certain adroitness in pursuit of their aim, one cannot say that they really refute the objections, nor that they have brought the contraction theory into satisfactory agreement with new research, especially in the field of geophysics. A thorough discussion of this neo-contraction theory must, however, be dispensed with here.

the earth's evolution in the form of the contraction theory. It is quite understandable that it is difficult for many European scientists to free themselves completely from this tradition and that they view the results of geophysics with a mistrust that never completely fades.

However, where does the truth lie? The earth at any one time can only have had one configuration. Were there land bridges then, or were the continents separated by broad stretches of ocean, as today? It is impossible to deny the postulate of former land bridges if we do not want to abandon wholly the attempt to understand the evolution of life on earth. But it is also impossible to overlook the grounds on which the exponents of permanence deny the existence of sunken intermediate continents. There clearly remains but one possibility: there must be a hidden error in the assumptions alleged to be obvious.

This is the starting point of displacement or drift theory. The basic "obvious" supposition common to both land-bridge and permanence theory—that the relative position of the continents, disregarding their variable shallow-water cover, has never altered—must be wrong. The continents must have shifted. South America must have lain alongside Africa and formed a unified block which was split in two in the Cretaceous; the two parts must then have become increasingly separated over a period of millions of years like pieces of a cracked ice floe in water. The edges of these two blocks are even today strikingly congruent. Not only does the large rectangular bend formed by the Brazilian coast at Cape São Roque mate exactly with the bend in the African coast at the Cameroons, but also south of these two corresponding points every projection on the Brazilian side matches a congruent bay on the African, and conversely. A pair of compasses and a globe will show that the sizes are precisely commensurate.

In the same way, North America at one time lay alongside Europe and formed a coherent block with it and Greenland, at least from Newfoundland and Ireland northwards. This block was first broken up in the later Tertiary, and in the north as late as the Quaternary, by a forked rift at Greenland, the sub-blocks then drifting away from each other. Antarctica, Australia and India up to the beginning of the Jurassic lay alongside southern Africa and formed together with it and South America a single large continent, partly covered by shallow water. This block split off into separate blocks in the course of the Jurassic, Cretaceous and Tertiary, and the sub-blocks drifted

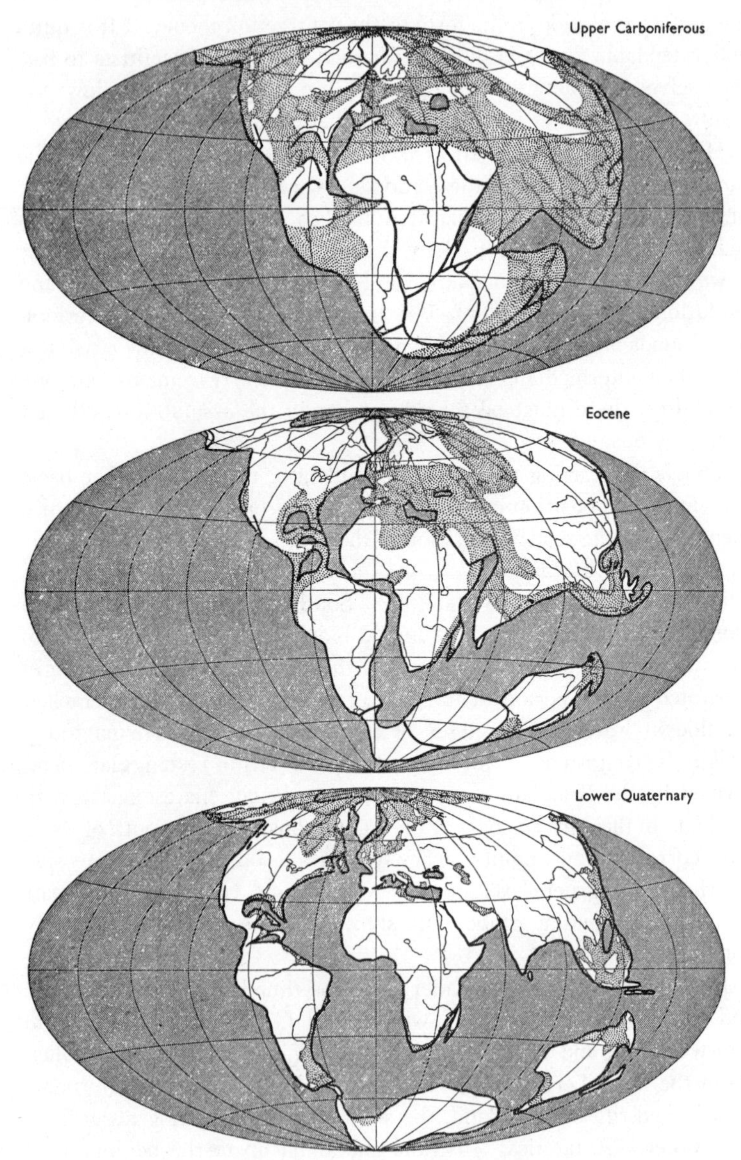


FIG. 4. Reconstruction of the map of the world according to drift theory for three epochs.

Hatching denotes oceans, dotted areas are shallow seas; present-day outlines and rivers are given simply to aid identification. The map grid is arbitrary (present-day Africa as reference area; see Chapter 8).

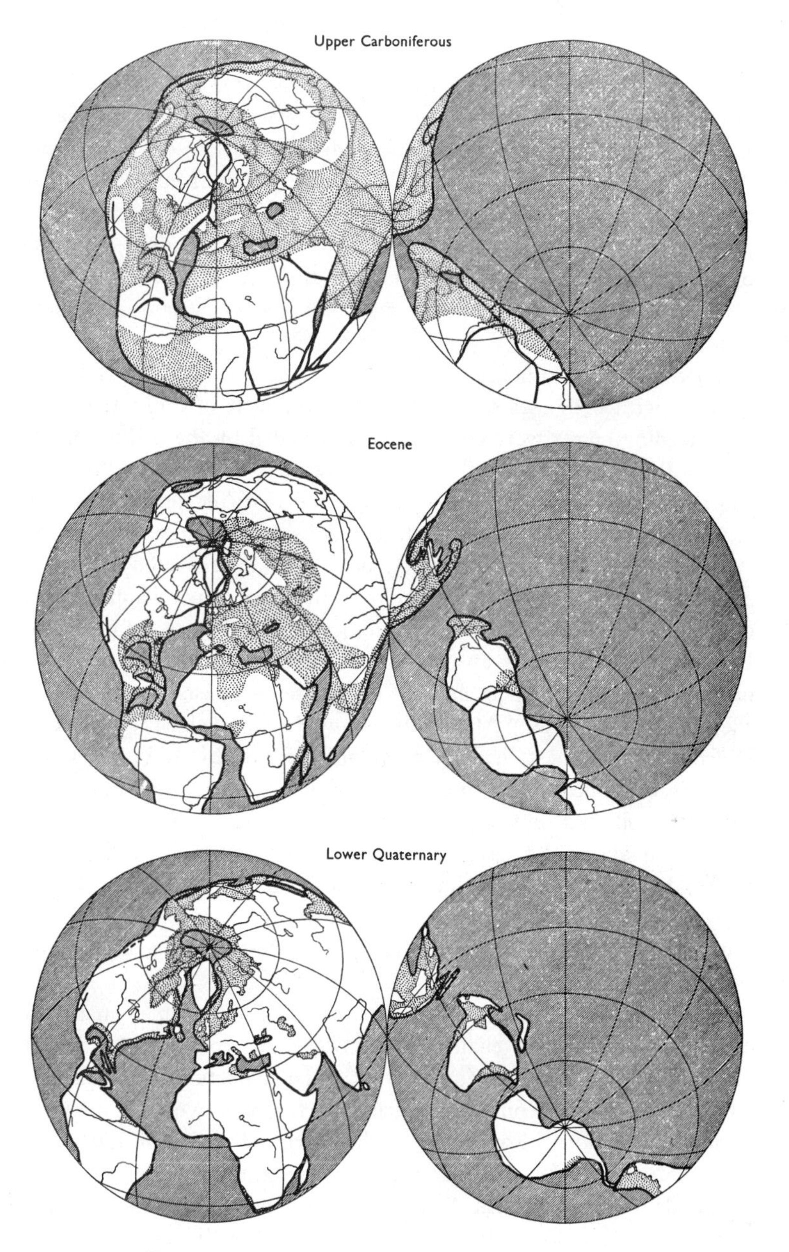


FIG. 5. Same as Fig. 4, in different projection.

away in all directions. Our three world maps (Figs. 4 and 5) for the Upper Carboniferous, Eocene and Lower Quaternary show this evolutionary process. In the case of India the process was somewhat different: originally it was joined to Asia by a long stretch of land, mostly under shallow water. After the separation of India from Australia on the one hand (in the early Jurassic) and from Madagascar on the other (at the transition from Tertiary to Cretaceous), this long junction zone became increasingly folded by the continuing approach of present-day India to Asia; it is now the largest folded range on earth, i.e., the Himalaya and the many other folded chains of upland Asia.

There are also other areas where the continental drift is linked causally with orogenesis. In the westward drift of both Americas, their leading edges were compressed and folded by the frontal resistance of the ancient Pacific floor, which was deeply chilled and hence a source of viscous drag. The result was the vast Andean range which extends from Alaska to Antarctica. Consider also the case of the Australian block, including New Guinea, which is separated only by a shelf sea: on the leading side, relative to the direction of displacement, one finds the high-altitude New Guinea range, a recent formation. Before this block split away from Antarctica, its direction was a different one, as our maps show. The present-day east coastline was then the leading side. At that time New Zealand, which was directly in front of this coast, had its mountains formed by folding. Later as a result of the change in direction of displacement, the mountains were cut off and left behind as island chains. The present-day cordilleran system of eastern Australia was formed in still earlier times; it arose at the same time as the earlier folds in South and North America, which formed the basis of the Andes (pre-cordilleras), at the leading edge of the continental blocks, then drifting as a whole before dividing.

We have just mentioned the separation of the former marginal chain, later the island chain of New Zealand, from the Australian block. This leads us to another point: smaller portions of blocks are left behind during continental drift, particularly when it is in a westerly direction. For instance, the marginal chains of East Asia split off as island arcs, the Lesser and Greater Antilles were left behind by the drift of the Central American block, and so was the so-called Southern Antilles arc (South Shetlands) between Tierra del Fuego and western Antarctica. In fact, all blocks which taper off towards the south exhibit a bend in the taper in an easterly direction because the tip has trailed behind: examples are the southern tip of Greenland, the Florida shelf, Tierra del Fuego, the Graham Coast and the continental fragment Ceylon.

It is easy to see that the whole idea of drift theory starts out from the supposition that deep-sea floors and continents consist of different materials and are, as it were, different layers of the earth's structure. The outermost layer, represented by the continental blocks, does not cover the whole earth's surface, or it may be truer to say that it no longer does so. The ocean floors represent the free surface of the next layer inwards, which is also assumed to run under the blocks. This is the geophysical aspect of drift theory.

If drift theory is taken as the basis, we can satisfy all the legitimate requirements of the land-bridge theory and of permanence theory. This now amounts to saying that there were land connections, but formed by contact between blocks now separated, not by intermediate continents which later sank; there is permanence, but of the area of ocean and area of continent as a whole, but not of individual oceans or continents.

Detailed substantiation of this new concept will form the chief part of the book.

CHAPTER 3

# *Geodetic Arguments*

WE BEGIN the demonstration of our theory with the detection of present-day drift of the continents by repeated astronomical position-finding, because only recently this method furnished the first real proof of the present-day displacement of Greenland—predicted by drift theory—and because it also constitutes a good quantitative corroboration. The majority of scientists will probably consider this to be the most precise and reliable test of the theory.

Compared with all other theories of similarly wide scope, drift theory has the great advantage that it can be tested by accurate astronomical position-finding. If continental displacement was operative for so long a time, it is probable that the process is still continuing, and it is just a question of whether the rate of movement is enough to be revealed by our astronomical measurements in a reasonable period of time.

To assess the question, we must go a little more deeply into the matter of absolute time scales of the geological periods. Evaluation of these is known to be doubtful, but not to the extent that no answer to our question is possible.

The time which has elapsed since the last ice age has been estimated by A. Penck on the basis of his studies of Alpine glaciation as 50,000 years; by Steinmann as at least 20,000, at most 50,000; by Heim, on the basis of calculations carried out in Switzerland, and also by United States glacial geologists, as only about 10,000 years. By means of astronomical investigations Milankovitch arrived at a climatic high point for the last ice age at about 25,000 years ago (the main phase of this ice age took place 75,000 years ago) and an immediately following climatic optimum at about 10,000 years ago; this optimum has been confirmed in northern Europe by geological evidence. De Geer concluded from his loam-strata counts that the retreating icecap edge passed Skåne in southern Sweden 14,000 years ago, but was still at Mecklenburg 16,000 years ago. By Milankovitch's reckoning, the duration of the Quaternary was 0.6 to 1 million years. For our purposes the agreement between these estimates is good enough.

Attempts have been made to assess the duration of earlier epochs by measuring the thickness of their sediment deposits. Dacqué [171] and Rudzki [170], for example, have used this method to deduce that the length of the Tertiary was of the order of 1 to 10 million years. The Mesozoic was about three times as long, the Palæozoic about twelve times.

Very much longer time scales, particularly for the earlier periods, are provided by radioactive dating, which enjoys the greatest esteem today [207]. The procedure is based on the gradual decay of uranium and thorium atoms, which emit alpha particles (helium nuclei) and finally become lead atoms after undergoing several intermediate transformations.

Three different methods of this type of dating can be distinguished. The first is the helium method, in which the relative amount of helium produced in increasing concentration in the mineral is measured. The numbers provided are smaller than those in the following method, or so it is thought, because some helium slowly escapes, and the He method is held to be inferior. Another way is to determine the relative amount of the end product, the lead, and to deduce the date from this. The third method is that of "pleochroic haloes," which arise because alpha emission produces a very small coloured halo around the radioactive substance in the rock. In the course of time, the halo expands and the sample can be dated by the size of the halo.

Born (in Gutenberg [45]) has used this method to date a Miocene rock as $6 \cdot 10^6$ years old, a Miocene–Eocene rock as $25 \cdot 10^6$ and a late Carboniferous rock as $137 \cdot 10^6$. These three values were obtained by the He method. The Pb method gave appreciably higher values—$320 \cdot 10^6$ for the late Carboniferous and as much as $1200 \cdot 10^6$ for the Algonkian (late Proterozoic), where the He method gave only $350 \cdot 10^6$ years. These values are very much greater than the estimates based on thickness of sediment deposits. [6]

[6] Although it is undoubtedly true that geological periods are longer the older they are, Dacqué's view [171] still does not seem quite justified to me, when he

However, since we are dealing here in the main only with post-Tertiary times, where the various methods yield passably similar data, the results are sufficient for our purpose. We can therefore take the following figures as a basis:

| | |
|---|---|
| Time elapsed since beginning of Tertiary times | 20 million years |
| ,, ,, ,, ,, ,, Eocene ,, | 15 ,, ,, |
| ,, ,, ,, ,, ,, Oligocene ,, | 10 ,, ,, |
| ,, ,, ,, ,, ,, Miocene ,, | 6 ,, ,, |
| ,, ,, ,, ,, ,, Pliocene ,, | 3 ,, ,, |
| ,, ,, ,, ,, ,, Quaternary | 1 ,, ,, |
| ,, ,, ,, ,, ,, post-Quaternary | 10–50,000 ,, |

With the help of these figures and the distances covered by the continents we can form a rough picture of the amount of annual drift, assuming that the displacements took place and are still taking place at a uniform speed. These two assumptions are of course rather difficult to test. If we add the fact that uncertainty in dating may be as great as 50%, perhaps even 100%, and further that disruption dates are uncertain, it will be immediately clear that the figures which follow can only serve to provide a rough orientation, and that no one should be surprised if later measurements give quite different results. In spite of all this, the rough calculations are very useful in that they draw attention to areas where there is some prospect of being able to measure displacement over a shorter time span.

The table on page 26 gives the expected annual increase in separation for a number of specially interesting areas.

The greatest change is thus to be expected for the gap between Greenland and Europe, then for that between Iceland and Europe and between Madagascar and Africa. In the case of Greenland and Iceland the drift lies in an east–west direction, so that astronomical position-finding can detect only increases in the longitudinal difference, not the difference in latitude.

As it happens, attention was already called some time ago to the increase in the longitudinal differential between Greenland and Europe. The history of this discovery is not wholly without interest.

---

says that such vast time spans for the earlier periods are contradicted by thicknesses of sedimentary deposits, and when he cites this as a reason for distrusting radioactive dating. This question does not come up, however, in the case of the more recent geological periods, which are the only ones considered here.

At the time when I had worked out a first rough sketch of the drift theory, the longitude determinations made by the Danmark expedition to northeastern Greenland had still not been computed. (This was the 1906–1908 expedition, led by Mylius-Erichsen, and I had taken part in it as assistant.) However, it was well known to me that there were earlier determinations on hand from the working area of our expedition, and that the relation between these earlier data

| Area | Relative drift to date (km) | Period since separation, approx. ($10^6$ years) | Displacement per year, average (m) |
|---|---|---|---|
| Sabine Island–Bear Island | 1070 | 0.05–0.1 | 21–11 |
| Cape Farewell–Scotland | 1780 | 0.05–0.1 | 36–18 |
| Iceland–Norway | 920 | 0.05–0.1 | 18–9 |
| Newfoundland–Ireland | 2410 | 2–4 | 1.2–0.6 |
| Buenos Aires–Cape Town | 6220 | 30 | 0.2 |
| Madagascar–Africa | 890 | 0.1 | 9 |
| India–southern Africa | 5550 | 20 | 0.3 |
| Tasmania–Wilkes Land | 2890 | 10 | 0.3 |

from the Sabine Island longitude stations and our data taken at Danmarkshavn had been established by triangulation. I therefore wrote to the expedition's cartographer (J. P. Koch), gave him an outline of my drift theory and asked him whether the longitude data from our expedition had diverged from the earlier figures in the expected way. Koch made a provisional calculation from the data and informed me that in fact a difference of the expected order of magnitude was shown, but that he was not able to credit the difference to a displacement of Greenland. When a definitive computation was made, Koch investigated the sources of error with special reference to this question and came to the conclusion this time that drift theory was in fact the most plausible explanation [172]: "From what has gone before, it would appear that the sources of error, whether taken individually or all together, are insufficient to explain the difference of 1190 m between the position of Haystack according to the data of the Danmark expedition and those of the Germania expedition (1869–1870). The sole source of error to be considered here is that of the astronomical longitude determination. However, to explain the divergence by incorrect specification of the site of the observatory we would have to assume actual errors in the astronomical determinations of longitude four to five times larger than their average error. . . ."

Sabine had already made longitudinal determinations in northeastern Greenland in 1823, so that there were altogether three sets of figures. It is true that these oldest measurements were not made in precisely the same place; Sabine made his observations on the southern edge of the island named after him, and unfortunately there is a degree of uncertainty, not very important, about the exact site of the observations, which was not marked. Börgen and Copeland made observations during the 1870 Germania expedition at the same place but a few hundred metres farther east; Koch's observations, on the other hand, were made far to the north at Danmarkshavn in Germania Land, but were connected to Sabine's by triangulation. The imprecision resulting from the transference of results from one point to another was examined accurately by Koch. The outcome was that this error was negligible compared with the much greater imprecision of the longitude determinations themselves. The data yield the following increase in the gap between northeastern Greenland and Europe:

1823–1870 – 420 m (= 9 m/yr.)
1870–1907 – 1190 m (=32 m/yr.)

The mean errors for the three series of observations were:

1823 – ca. 124 m
1870 – ca. 124 m
1907 – ca. 256 m

Now, of course, Burmeister [173] has rightly objected that in the case in point, which involved moon observations, the mean error cannot guarantee the accuracy of the results as it would in other cases. This is mainly because, in taking lunars, systematic errors may also arise which are not expressed in the mean error, and in unfavourable circumstances the systematic error may be as high as the value of the result itself or even exceed it. It was therefore possible to deduce from the observations only that they fit the hypothesis of drift very well, and are best explained thereby, but do not constitute accurate proof.

Since then, the Danish Survey (now the Geodetic Institute in Copenhagen) has taken up the question in a gratifying fashion. P. F. Jensen [174] carried out new longitude determinations in western Greenland during the summer of 1922 with this in mind, using the far more precise method of radio telegraphy time transmissions. A. Wegener [175] and Stück [176] have reported in German on his results. Jensen carried out two operations in Greenland. One was to repeat the earlier longitude determinations in the Godthaab Colony, in order to make a comparison with these older observations, which date partly from 1863 (made by Falbe and Bluhme) and partly from the International Polar Year 1882/1883 (made by Ryder). These older observations were made as lunars and were correspondingly imprecise. Jensen therefore combined them into an average measurement corresponding to the year 1873 and compared this with his own much more accurate measurements, which above all were free of any chance of large systematic errors. Again the upshot was that in the interim period Greenland had drifted about 980 m west, the equivalent of 20 m/yr.

I have presented [175] the results of these measurements together with those of the eastern Greenland observations in Figure 6 to help

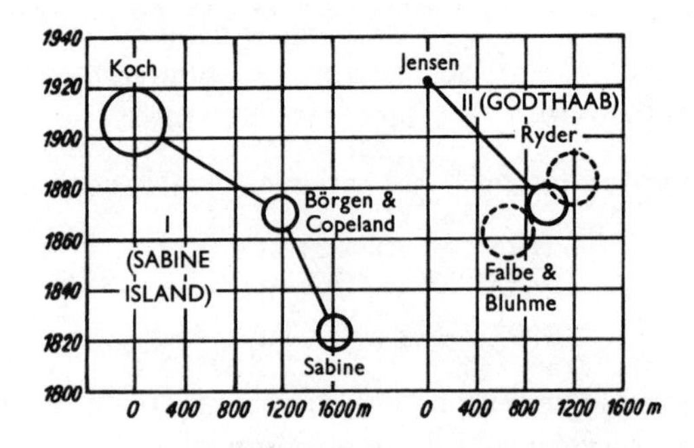


FIG. 6. Displacement of Greenland according to earlier determinations of longitude.

the reader visualise the situation. The radii of the circles, which can be read off by means of the scale on the abscissa, equal the mean error of the series of measurements expressed in meters. The superior accuracy of Jensen's observations is immediately evident. Observations grouped under "I" refer to the Sabine Island (northeastern Greenland) data, those under "II" to data taken at Godthaab in western Greenland. Besides the average of the older observations mentioned above, the 1863 and 1882/1883 data are also shown; their vector difference, of course, is in the counter direction, but as the time interval is so short one should not read into this more than the effect of their imprecision. However, each of them gives a time rate of longitude increase when compared with Jensen's much later observations. All in all, therefore, there now were the following four mutually independent sets of comparative data:

Koch–Börgen and Copeland,
Koch–Sabine,
Jensen–Falbe and Bluhme,
Jensen–Ryder,

which all corresponded with drift theory. While it is true that all these comparisons suffer wholly or partly from the fact that they are based on lunars and *may* therefore contain systematic errors that cannot be checked, this accumulation of similar results which do not stand in opposition to any others makes it highly improbable that it is all just a matter of an unfortunate combination of extreme errors of observation.

The Danish Survey Organisation, however, fortunately undertook to repeat these longitude determinations at regular intervals as part of its programme. In accordance with this, Jensen's second operation consisted of installing a suitable observatory in Kornok, in the favourable climate of the upper section of Godthaab Fjord, and carrying out the first standard determination of longitude with the aid of precise radio time transmissions. In 1922 he measured the longitude of Kornok as:

3 h 24 m 22.5 s ±0.1 s west of Greenwich (stellar observations),
3 24 22.5 ±0.1 " " " (solar observations).

The determination of the longitude of Kornok has now been repeated (summer, 1927) by Lieutenant Sabel-Jörgensen [209], using the modern impersonal micrometer which eliminates the "personal equation." This allows far greater accuracy to be achieved than was possible in Jensen's measurements.

The result, which was awaited with considerable interest, was: 1927 longitude of Kornok: 3 h 24 m 23.405 s ± 0.008 s.[7]

*Comparison with Jensen's figure yields an increase in the longitude difference relative to Greenwich, i.e., in the distance of Greenland from Europe, of about* 0.9 *seconds (time) in five years, or about a rate of* 36 *m/yr.*

[7] The author is most grateful to the Director of the Geodetic Institute in Copenhagen, Professor Nörlund, for his permission to quote this as yet unpublished result.

This increase is nine times larger than the mean error of observations, and there is no question of any systematic error in radio telegraphic time transmissions. *The result is therefore proof of a displacement of Greenland that is still in progress*, unless it should be supposed that Jensen's "personal equation" amounted to 0.9 seconds of time—a most improbable hypothesis.

The Kornok measurements are to be repeated every five years by the impersonal method. It will be of interest to determine the annual displacement quantitatively with greater accuracy and to establish whether the drift rate is steady or variable.

As a result of this first precise astronomical proof of a continental drift, which fully corroborates the predictions of drift theory in a quantitative manner, the whole discussion of the theory, in my view, is put on a new footing: interest is now transferred from the question of the basic soundness of the theory to that of the correctness or elaboration of its individual assertions.

As our table shows, circumstances are less favourable for measurement of the relative displacement rate of North America and Europe than was the case for Greenland. Of course, conditions are more favourable in that we are not dependent on lunars, because even the earlier longitude readings in North America were already taken by means of the telegraphic cable. The price we have to pay for this advantage is that the expected change is extremely small. Our table gives about 1 m/yr, but this is the average since the connection between Newfoundland and Ireland was severed. Since then, however, a change in direction of North America must have occurred as a result of the breakaway of Greenland, which is still in progress; it is probable that North America has since been drifting more towards the south relative to the substrate. This also appears to follow from the present-day relative position of the corresponding coastal points of Labrador and southwestern Greenland, and is further corroborated by the rift line of the San Francisco seismic fault and the incipient compression of the Californian peninsula. It is therefore hard to say how large the expected increase in longitude will prove to be; in any case, it should be somewhat less than the calculated value of 1 m/yr.

Using as a basis the older longitude measurements made by transatlantic cable in 1866, 1870 and 1892, I once deduced that the actual increase in separation was as much as 4 m/yr. According to Galle [177], however, this result must be due to an unfavourable combination of measurement data. The difficulty here is fundamental because the older measurements do not refer to the same places in Europe and North America, so that one has still to take into account the longitude differences within the continents; somewhat different results are obtained for these according to the method used and this affects the result. Shortly before the [First] World War, a new longitude determination was in progress in cooperation with America, in order to decide this question; radio telegraphy was employed to check the measurements. Although the measurements were prematurely disrupted at the onset of war when the cable was cut, and the result thus does not have the desired accuracy, it appears nevertheless that the shift is still too small to be reliably demonstrated at the present time. The figures for the longitudinal difference of Cambridge and Greenwich were found to be [178]:

| | | | | |
|---|---|---|---|---|
| 1872 | – | 4 h | 44 m | 31.016 s |
| 1892 | – | 4 | 44 | 31.032 |
| 1914 | – | 4 | 44 | 31.039 |

The oldest determination, which I evaluated as 4 h 44 m 30.89 s, is omitted from the list, as it is alleged to be too inaccurate.

Since 1921, successive determinations of longitudinal difference between Europe and North America have been carried out by radio time signals; Wanach [179] has discussed these results up to 1925. Since only four years are involved, it is not surprising that no clear increase can be detected as yet. However, even these observations in no way preclude such an increase: quite the contrary, if combined they give an annual westward shift of America of 0.6 m, though the possible error is $\pm 2.4$ m. Wanach concludes: "At the moment one can only say that any displacement of America with respect to Europe of appreciably more than 1 m/yr is most improbable." Brennecke [229] arrived at a similar verdict: "The data obtained are not, it is true, evidence in favour of drift of the continent by the amount stated, but neither are they evidence against it. We must still await the outcome." It should be noted here that the older determinations made by transatlantic cable are entirely disregarded when making the new observations by radio. This is justified in so far as the cable observations are appreciably less precise than radio measurements. It may be, however, that this deficiency is compensated by the much greater time intervals that were available then, and it would thus be worth while to combine the old observations with the new. This must be left to the geodesists. I have no doubt, however, that in the not too distant future we will be successful in making a precise measurement of the drift of North America relative to Europe.

Attention has also been drawn recently to the change in the geographical coordinates of Madagascar. The longitude of the Tananarive observatory was found in 1890 with the help of lunar culminations, and after its destruction and restoration, at the same spot measurements were made in 1922 and 1925 by the radio telegraphy method [180]. I am grateful to Prof. C. Maurain (Paris) for his letter which gives the three positions:

| Year | Observer | Method | Longitude east of Greenwich |
|---|---|---|---|
| 1889–1891 | P. Colin | Lunar culmination | 3h 10m 7s |
| 1922 | P. Colin | Radio telegraphy | 3 10 13 |
| 1925 | P. Poisson | Radio telegraphy | 3 10 12.4 |

These values indicate a shift of Madagascar relative to the Greenwich meridian of 60–70 m/yr, a large amount. In our table on page 26 the shift relative to Africa is assigned a much smaller figure. It therefore appears as though southern Africa, too, is moving in an easterly direction relative to Greenwich; drift theory can make no further useful pronouncements on this because of the large separation of these areas from each other. It is to be hoped that the longitudes of southern Africa will also be surveyed in the future, so that the longitudinal difference between Madagascar and southern Africa, a matter of the greatest importance in drift theory, can also be monitored. Repeated accurate latitude measurements for both areas would also be necessary, so that the other component of relative movement between Madagascar and Africa can be followed quantitatively. In any case, however, the observed longitudinal change of Madagascar is in the right direction to fit drift theory. It should also be noted here, of course, that the oldest measurements were based on lunars and that the same objections can be raised against them as against the northeastern Greenland data. Nevertheless, the overall shift, almost $2\frac{1}{2}$ km, is so large that any idea that it is just due to errors in observation has little claim to probability. However, provision has been made on Madagascar, too, for further repetitions of the measurements, so that before long we can expect reliable measurements to reach us from there also.

At the Congress on Geodesy in Madrid in 1924 and at the Conference of the International Astronomical Union in 1925, a comprehensive plan for following continental drift by radio telegraphic determinations of longitude was drawn up. According to this, measurements are to be made not only between Europe and North America, but also in Honolulu, eastern Asia, Australia and Indochina. The first series of measurements in this program was carried out in the fall of 1926; G. Ferrié [213] has just reported on the results obtained by the French. Any possible changes would naturally only show up after later repetitions. It would seem that little consideration has been given in the plan to deciding which parts of the world should be expected to show measurable changes based on drift theory. The examples of Greenland and Madagascar, however, allow us to hope that the plan is being improved in this direction.

At all events, it should be clear that exact testing of drift theory by repeated astronomical position-finding is already in progress to a large extent and that the first indications that the theory is right have already been found.

In conclusion, it should be remembered that latitudinal changes have been noted for some time now by European and North American observatories.

As Günther reports [181], A. Hall regarded the following *decreases* in latitude as certain:

| | | | |
|---|---|---|---|
| Paris | — 1.3″ | in 28 | yr. |
| Milan | — 1.51″ | ,, 60 | ,, |
| Rome | — 0.17″ | ,, 56 | ,, |
| Naples | — 1.21″ | ,, 51 | ,, |
| Königsberg (Prussia) | — 0.15″ | ,, 23 | ,, |
| Greenwich | — 0.51″ | ,, 18 | ,, |

Kostinsky and Sokolow stated that Pulkowa (Pulkovo) records a centennial decrease in latitude. In addition, there is the decrease at Washington of 0.47″ in 18 years.

It was discovered that systematic errors of similar magnitude can arise due to so-called "chamber refraction" in the observatory dome. As a result, there was a long-standing tendency to attribute all variations to this source of error.

Nevertheless, opinions have multiplied recently in favour of regarding such changes as real, particularly since Lambert [182] showed that the latitudes of Ukiah in California and of other North American stations are apparently changing at the present time. In a more recent work [221] Lambert states: "The international stations are not the only ones where perplexing changes in latitude have occurred. Rome has apparently changed its latitude by 1.43″ since 1885. A systematic study of such anomalies would be highly desirable."

However, the striking fact is that the present-day shift is in the opposite sense to those older ones mentioned above, because the latitude of Ukiah is increasing.

It is hard to interpret these latitudinal shifts because they could be caused by continental drift as well as by polar wandering, and the latter does not have to be connected with the former. As we will show in more detail later, it has been possible in very recent times to detect present-day polar wandering by means of the measurements of the International Latitude Service, according to which the North Pole is shifting in the direction of North America; this means an increase of latitude for the North American stations. However, the extent of this polar shift is less than the observed North American increase in latitude, judging by results to date. If the polar shift should not turn out in the future to be any larger, one would conclude that North America is drifting northwards relative to the rest of the earth's surface: this would be very odd, because there are many indications that it is drifting southwards. The complete interpretation of these matters will only be possible after a more prolonged series of observations has been made, and whether a clear understanding of the older shifts will ever be reached seems doubtful in the circumstances.

## CHAPTER 4

# *Geophysical Arguments*

THE statistical distribution, taken over the earth's crust, of heights above and below sea level leads to the remarkable conclusion that there are two modal values of elevation, while intermediate values are rare. The higher value corresponds to the elevation of the continental tables, the lower to the ocean floors. If one imagines the whole surface of the earth subdivided into squares of 1-km side and arranged

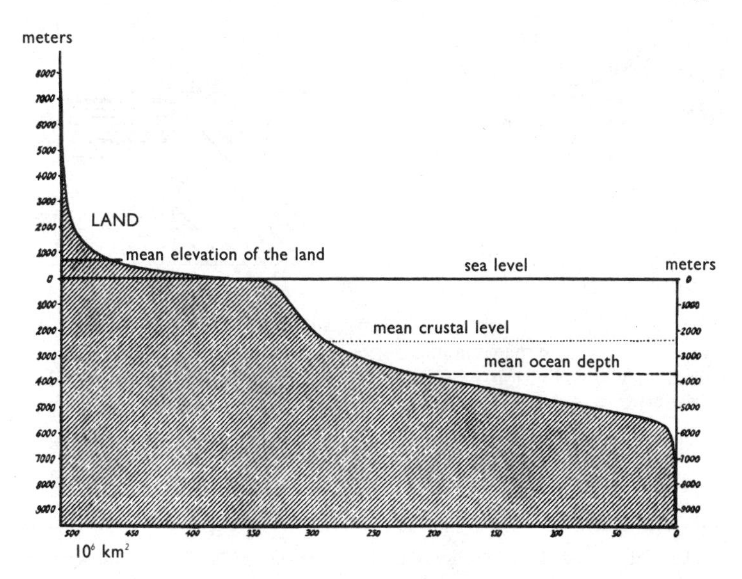


FIG. 7. Hypsometric curve of the earth's surface (according to Krümmel).

in order of height above and below sea level, the result is the well-known hypsometric curve of the earth's surface (Fig. 7), which clearly exhibits two steps. H. Wagner's calculations [28] show that numerically the frequencies of the various data groups come out as follows:[8]

| Depths | | | | | | | Heights | | | | |
|---|---|---|---|---|---|---|---|---|---|---|---|
| <6 | 5–6 | 4–5 | 3–4 | 2–3 | 1–2 | 0–1 | 0–1 | 1–2 | 2–3 | >3 | km |
| 1.0 | 16.5 | 23.3 | 13.9 | 4.7 | 2.9 | 8.5 | 21.3 | 4.7 | 2.0 | 1.2 | % |

This series is best represented in another form drawn up by Trabert [31] on the basis of somewhat earlier data, which differ only trivially. Figure 8 shows this frequency distribution; it refers to 100-m increments, so that the percentages are obviously only about one-tenth of those in the series given above. The two maxima here are at ca. 4700-m depth and ca. 100-m height relative to sea level.

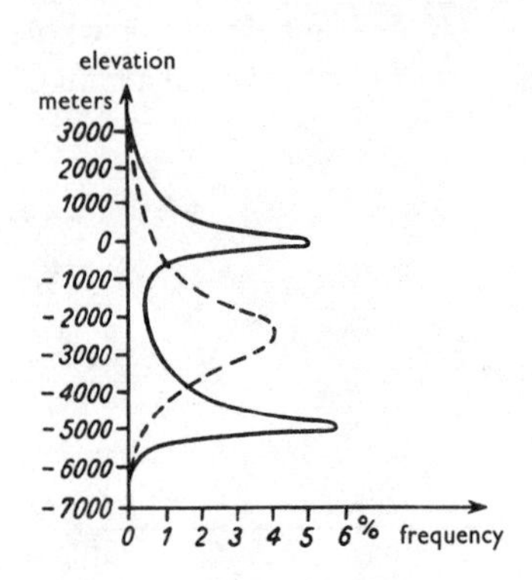


FIG. 8. The two maxima in the frequency distribution of elevations.

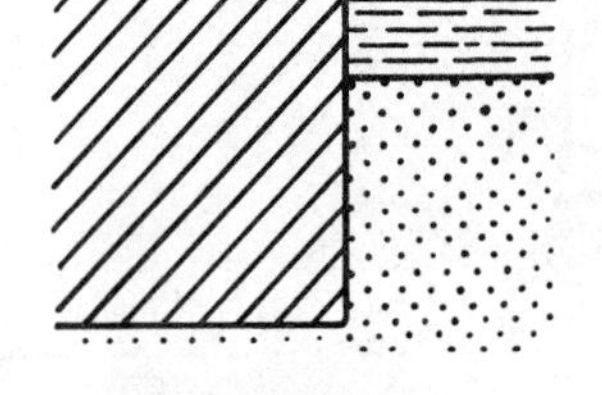

FIG. 9. Diagrammatic cross section through a continental margin.
Horizontal hatching = water.

In considering these figures it should be noted that as the number of soundings increases, the abrupt drop from the edge of a continent or

[8] These figures are based on Kossinna's oceanic survey [29]. Our illustrations are drawn from the older and slightly different data of Krümmel [30] and Trabert [31].

shelf to the ocean appears constantly steeper, as any comparison of earlier ocean charts with the more recent ones of Groll [32] will show. For example, as late as 1911 Trabert gave 4% for the 1–2 km group and 6.5% for the 2–3 km, but we now find that Wagner, whose data are ultimately based on the Groll charts, gives only 2.9 and 4.7%, respectively, for the same groups. One can therefore probably expect that in the future the two frequency maxima will appear even more sharply differentiated than they do from observations to date.

In the whole of geophysics there is probably hardly another law of such clarity and reliability as this—that there are two preferential levels for the world's surface which occur in alternation side by side and are represented by the continents and the ocean floors, respectively. It is therefore very surprising that scarcely anyone has tried to explain this law, which has, after all, been well known for some time. If, in fact, according to the usual geological interpretation, the elevations resulted from uplift and the depths from subsidence of but one uniform original level, and if (as seems obvious) the levels are less frequent, the greater their height or depth relative to sea level: then the resulting frequency distribution would have to approximate the Gaussian error curve (roughly, the broken-line curve in Fig. 8). Had this been so, there would have been only a single peak in the distribution, corresponding approximately to the mean crustal level (−2450 m). Instead, we observe two peaks, and for each one the curve on either side follows roughly the course of the error function. From this we conclude that there were at one time two undisturbed primal levels, and it seems an inevitable deduction that we are dealing with two different layers in the crust when we refer to the continents and the oceans. To put it in rather picturesque terms, the two layers behave like open water and large ice floes. Figure 9 gives a diagrammatic cross section through a continental margin according to this new concept.

In this way we have achieved for the first time a plausible explanation of the old question of the relationship between the great ocean floors and the continental blocks. As far back as 1878 A. Heim [33] touched on this problem when he said "that, until more precise observations have been made in connection with prehistoric continental displacements, . . . and until we have more complete data on the extent of the compensatory compressions that formed most mountain ranges, we can hardly expect any really reliable progress in our understanding of the original interrelationship between mountains and continents or of the configuration of the latter relative to each other."

The problem proved to be more pressing the more numerous the bathymetric soundings became and the sharper the contrast they revealed between the broad, flat expanses of the ocean floors and the continental areas, likewise flat but set about 5 km higher. In 1918 E. Kayser [34] wrote: "Compared with the volume of these vast rock formations (the continental blocks), all mainland upthrusts appear minor and insignificant. Even high ranges such as the Himalaya are merely tiny wrinkles on the surface of that supporting base. This fact alone makes the older viewpoint, in which mountains are supposed to represent the fundamental framework of the continents, seem untenable today. . . . We should rather take the opposite view, that the continents are the earlier, determinative formations, and the mountains only subsidiary and more recent."

The solution to this problem offered by drift theory is so simple and obvious that one can hardly believe it could occasion dissent. Nevertheless, some opponents of the theory have tried to explain the double peak in the level frequency distribution by some other method. Yet all these attempts have miscarried. Soergel [35] believed that if, starting from a given level, part of one side were raised and on the other side a part were lowered, the intermediate section thus being much reduced by making the sides sloping, two frequency maxima would result, corresponding to the raised and sunken parts. Similarly, G. V. and A. V. Douglas [36] thought that if the original level section were transformed by folding into a sinusoidal surface, two maxima should be formed, corresponding to wave crest and trough. Both views are derived from the same fundamental mistake, that of confusing individual process with statistical resultant. In the latter, the geometry of the former is of no concern whatever. It is simply a matter of whether, in the infinite series of elevations and depressions (to use Soergel's terminology) or of folds (to use Douglas's), it is possible for two maxima to appear in the frequency distribution *and for the numerical values of the individual levels still to vary arbitrarily.* Obviously, this could only be true if some tendency towards selection of preferential levels were operative. Yet this is not the case. For uplift and subsidence just as for fold elevations, we know only one law: the greater, the rarer. The highest frequency must therefore always fall to the share of the original level, and the frequencies of levels above or below the original must decrease roughly in accordance with the Gaussian error function.

It should also be recalled here that some authors, Trabert [31] in particular, have put forward the view that the ocean floors may have been formed by more intensive cooling of the underlying rock by the cold ocean waters. However, it emerges from Trabert's own calculations that one would have to assume that the cooling of the ocean-floor sectors extended to the centre of the earth. Since that appears inadmissible, his computations would be better suited to confuting than to proving the hypothesis. Besides, it is easy to see that by his method one could only deduce a general tendency for already existing depressions in the earth's surface to become deeper; but one could not use it to explain the existence of a floor set at almost the same depth in every ocean, that is, of the second peak in the frequency distribution of contours. This point was recently emphasised by Nansen [222]. In fact, of course, this explanation, which originated with Faye, is less and less often referred to nowadays, especially since the basis for assessing the earth's heat balance has completely changed since the discovery of radium in the crust.

Naturally, I must immediately caution the reader against carrying the new concept of the nature of ocean floors too far. Even in our comparison with the tabular icebergs made earlier, one must bear in mind that the surface of the sea between them can again be covered with fresh ice, and also that the water can be covered over once more with smaller fragments of the iceberg, calved from its upper edge or floating up from the foot of the berg far below the water line. Analogous processes will obviously also occur in many places on the ocean floors. Islands are usually larger fragments of continents, their substructures extending to about 50 km below the ocean floor (gravity measurements make this hypothesis probable). Further, it should be taken into account that the continental blocks, however brittle they may be at the surface, become plastic at considerable depths and can be deformed here as though they were dough; this means that when blocks separate, material therefrom (of correspondingly reduced thickness) can spread in this way, too, over small or extensive areas of the ocean floor. The Atlantic floor must be considered particularly "inhomogeneous" in this respect, as it is traversed longitudinally by the mid-Atlantic ridge, but the other ocean basins show similar structures, with their island arcs and submarine banks. We shall go further into the details of this in the chapter on ocean floors.

It is not inconceivable that as research progresses the model put forward here may turn out to represent the main features only, and that complications will have to be introduced if the true situation is to be correctly described. I myself [37], when examining statistically the first echo soundings made by the Americans over the North Atlantic, found that the main peak in the frequency distribution occurred for an appreciably greater depth, about 5000 m, and that a secondary peak could be observed for the 4400-m sounding. The second maximum would indicate a multilayer structure, and it will only be possible to judge its reality on the basis of the much more numerous soundings of the German "Meteor" expedition; no examination of these results for this purpose has yet been made.

Naturally, the question arises whether the idea of the basic difference between continental blocks and ocean basins, and of horizontal displacement of the former, agrees with the other results of geophysics, and whether geophysics can in turn provide corroboration of the ideas.

First of all, so far as isostasy (see above) is concerned, there is obviously excellent agreement with the whole concept of drift theory, but we cannot really obtain direct proof of the correctness of our theory by considering isostasy. We intend to examine all this more closely in what follows.

The physical basis for the theory of isostasy was derived from gravity measurements. The theory originated with Pratt; the word itself was coined later by Dutton in 1892. As far back as 1855 Pratt had discovered that the Himalaya Mountains did not exert the expected attractive force on the plumb bob. According to Kossmat, the northerly component of the plumb-line deflection at Kaliana in the Ganges plain, 56 miles from the foot of the range, was only 1 second of arc, whereas the attraction due to the mountains should give a deflection of 58 seconds of arc. Similarly, the deflection at Jalpaiguri was only 1 second, instead of 77. In accordance with this is the fact, universally acknowledged, that the gravitational field strength of large mountains does not differ from the ordinary value to the degree expected; mountain massifs appear to be compensated in this respect by some kind of subterranean mass deficit, as has been shown by the work of Airy, Faye, Helmert and others. The matter was also discussed by Kossmat in his very illuminating review [38]. It has also been shown that the gravitational force measured at the surface of the oceans has about its normal strength in spite of the obviously large mass deficit represented by the ocean basins. Earlier measurements made on islands admitted of various interpretations, it is true. But all doubt was removed when Hecker, following a suggestion of Mohn's, carried out gravity determinations on board ship, as well, by simultaneous readings of a mercury barometer and a hypsometer. A short while ago, the Dutch geodesist Vening-Meinesz [39] even succeeded in making use of the much more accurate pendulum method for carrying out measurements in a submarine, and the results of the first voyages fully corroborated Hecker's conclusion that, broadly speaking, isostatic conditions also apply over the oceans and that therefore the apparent mass deficit of the basins is compensated by a subterranean mass surplus of some kind.

In the course of time, different conjectures have been put forward as to the nature of these underground mass deficits or excesses. Pratt considered the crust to be a sort of dough which originally had uniform thickness everywhere, but which rose up in continental regions by some sort of loosening process and was compressed in the ocean regions. The higher the elevation of the surface above sea level, the less the density or specific weight of the crust, according to him. But below the so-called equilibrium level (ca. 120 km below sea level), all horizontal density differentials vanished (Fig. 10).

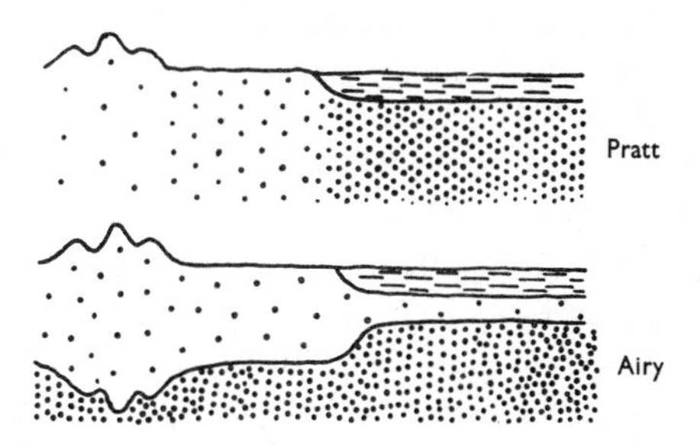


FIG. 10. Representations of isostasy according to Pratt and to Airy.

This idea was elaborated by Helmert and Hayford and employed as a general means of assessing observations on gravitational force. W. Bowie [224] is a principal exponent of this theory at the present time. He makes use of the following experiment as a means of explaining the idea: He floats on mercury a number of prisms consisting of materials of different density, such as copper, iron, zinc, pyrites, etc. The prisms should be of such a height that they all are immersed to the same depth in the mercury. Their common bottom surface then corresponds to the balancing or equilibrium level of the buoyancy force. Because of their unequal densities, the prisms project to various heights above the mercury meniscus, the densest material projecting the least and the lightest the most. This interpretation of the gravitational data finds a degree of support in the observed fact that in general the material of the earth's crust is less dense the higher above sea level it is found. Nevertheless, the idea that differences in density only extend to a certain definite depth, the equilibrium level, contains a physical improbability which can best be elucidated by means of Bowie's experiment. For the different prisms to have their undersides all at the same depth, their heights must bear a definite ratio to each other, determined by the ratios of their densities. If we divide up the earth's crust into prisms of different materials, then one and the same material must always and everywhere have a quite definite thickness, bearing a precise relationship to the thickness of each of the other materials, a relationship which is laid down once and for all on the basis of density. Yet there is no perceptible natural reason for such a connection between type of material (density) and thickness which would lead to the arbitrary condition of constant base level for all prisms.

Recently, many geodesists, such as Schweydar [40] and in particular Heiskanen [41, 42], have made use of another model with which to interpret gravitational data. This model, introduced as far back as 1859 by Airy, is also shown in Figure 10. Heim was probably the first to assume that the low-density crust under the mountain ranges is thickened and that the heavy magma on which it floats is pushed down to greater depths in these areas. Conversely, the low-density crust under deep-lying areas of the earth's surface, such as ocean basins, must be extra-thin. It is here assumed that there are only two types of material involved, a light crust and a heavy magma. Bowie illustrates this concept by an experiment analogous to the foregoing one: He floats a number of prisms of different heights but all of the same material (copper) on mercury. Clearly they all sink to different levels; the longest is immersed to the greatest depth but at the same time has the most elevated surface. It has often been stressed that this idea of Airy's fits much better than Pratt's the geological picture of the earth's crust, especially in the case of the large compressions that constitute fold mountains. On the other hand, Airy's conception leaves unexplained the twin peaks of the frequency distribution of the earth's contours, because it does not reveal why the light crust should turn out to have two basically different thicknesses, the thick continental blocks and the thin oceanic blocks.

The correct interpretation may be found in an amalgamation of both concepts. In the case of mountain ranges, we have to do basically with thickening of the light continental crust, in Airy's sense; but when we consider the transition from continental block to ocean floor, it is a matter of difference in type of material, in Pratt's sense.

Recent developments in isostasy theory deal mainly with the question of the range of its validity. For large blocks—for example, a whole continent or a whole ocean floor—the theory of isostasy must be accepted without qualification; but where smaller ones are concerned, such as individual mountains, the law loses its validity. Such small blocks can be supported by the elasticity of the whole block, just as a rock placed on an ice floe would be. Isostasy then operates between the floe and rock as a unit and the water. Thus continental gravitational measurements on geological structures with a diameter of hundreds of km very rarely show any deviation from isostasy; if the diameter amounts only to tens of km, there is generally only partial compensation; if it amounts to a few km only, there is almost no compensation at all.

Whether one bases one's view on the older concept of Pratt or on that of Airy and Heiskanen, examination of oceanic gravity measurements which reveal no sign of the large and visible mass deficit of the basins still leads to the result that the ocean floor consists of a denser, heavier material than that of the continental blocks. It is, of course, impossible to prove conclusively by this means that the greater density amounts not merely to a difference of physical state, but also to a difference of materials. However, rough calculations based on reasonable premises make this very probable.

The theory of isostasy does, however, provide a direct criterion for deciding the question of whether continents can drift horizontally. We have already referred above to the isostatic balancing movements, the finest example of which is the uplift of Scandinavia, still continuing today, at the rate of about 1 m per century. It can be regarded as an aftereffect of removal of the inland icecap, which completely melted more than 10,000 years ago, especially since the greatest present-day uplift is to be observed where the ice disappeared most recently. This is very elegantly demonstrated by the chart drawn up by Witting (according to Born [43]), which is reproduced as Figure 11.

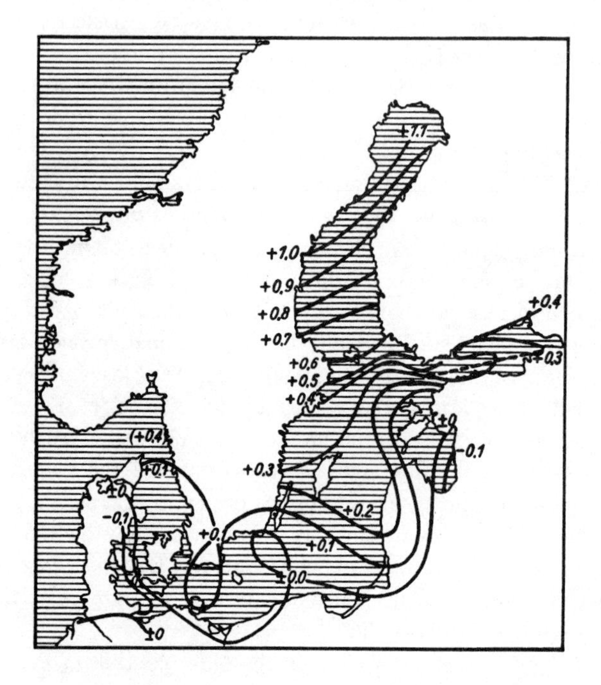


FIG. 11. Present-day uplift in the Baltic region indicated by tide-gauge measurements (cm/yr), according to Witting.

Born [43] has shown that this uplift region constitutes an anomaly in that the gravitational field is too weak, so far as can be judged from the as yet scanty observations (Fig. 12), and in fact this must be the case if the crust still lies below its equilibrium level. Nansen [222] has provided a particularly exhaustive description of all these phenomena related to the Scandinavian uplift; the greatest depression amounted to 284 m according to the high-water marks on the Ångermanland coast and probably 300 m inland. This uplift began slowly about 15,000 years ago, reached its highest rate—1 m in 10 years—7000 years ago and today is on the wane. The thickness of the central ice layer is estimated at about 2300 m. Vertical movement of such a large portion of crust obviously sets up flow in the substrate, so that displaced material is carried outwards. This is also borne out by the discovery made almost simultaneously by Born, Nansen, A. Penck, and Köppen (references in [43]), that the depression region of the inland icecap is surrounded by an annular zone of reduced uplift, to be attributed precisely to the substrate material forced outwards. At all events, the whole isostasy theory depends on the idea that the crustal underlayer has a certain degree of fluidity. But if this is so and the continental blocks really do float on a fluid, even

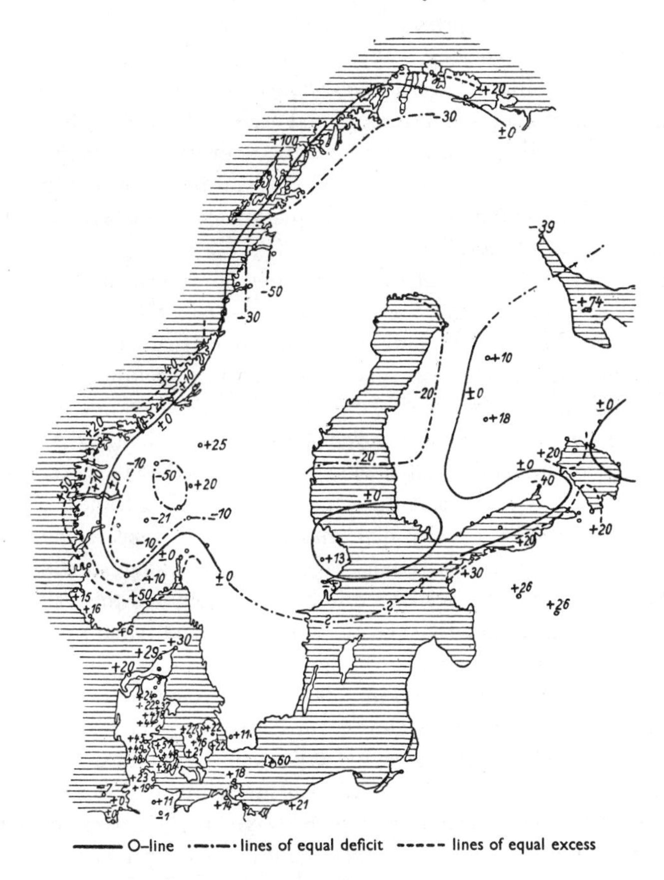


FIG. 12. Gravity anomaly in Scandinavia (according to Born).

though a very viscous one, there is clearly no reason why their movement should only occur vertically and not also horizontally, provided only that there are forces in existence which tend to displace continents, and that these forces last for geological epochs. That such forces do actually exist is proved by the orogenic compressions.

Of the greatest importance for our problems are the latest results of seismic research, which Gutenberg has in several places collected into a handy survey [44, 45].

Of the various types of seismic waves, it is well known that the longitudinal (primary) P waves and the transverse (secondary) S waves travel through the earth's interior ("preliminary" waves), while the L waves ("principal" waves) travel via the surface layer. The farther the recording station is from the focus, the greater the depth of penetration of the P and S waves that reach it. From the difference in time between tremour and arrival of waves at the station ("transit time") one can determine the velocity of propagation for the various depths. This velocity is a property of the material concerned and can therefore give information about the layer structure in the earth's interior.

It has been shown by such data that under Eurasia and also under the North American continental block, there is a very interesting boundary layer at about 50–60-km depth, where the velocity of the longitudinal waves jumps from 5.75 km/s (above) to 8.0 km/s (below), and that of transverse waves from 3.33 km/s (above) to 4.4 km/s (below). This boundary layer has till now been generally identified with the undersides of the continental blocks, as is suggested by the correspondence between its depth and the value of the block thickness derived by Heiskanen from gravity determinations.[9] It appears, however, that this interpretation cannot be upheld any longer; the block thickness must now be considered to be only about half this value, and the boundary layer described corresponds to an additional subdivision of the substratum. The layer is, however, completely absent under the Pacific Ocean. In this region one finds even for the surface layers a velocity of seismic waves almost equal to that given above for the zone below the boundary layer, i.e., 7 km/s for longitudinal waves and 3.8 km/s for transverse waves (whereas for continental surface layers these figures are 5.75 and 3.33 km/s, respectively). There can be but one interpretation possible for these figures, namely that the uppermost layers, which extend down to 60 km depth below the continental tables, are lacking in the Pacific region.

As was to be expected, the velocity of the surface waves, also a physical constant of the given material, also exhibited a corresponding

[9] On the basis of Pratt's theory greater block thicknesses (100–120 km) had been arrived at, while Airy's theory gives almost the same result as that of seismology. This counts in favour of the Airy concept, which has the recognised advantage in other respects also.

difference between values for ocean floors and continental blocks. We can regard this today as an established fact, since it was determined independently by five different research workers. In 1921, Tams [46] found the following velocities for surface waves, taken from a selection of specially clear recordings:

| 1. Oceanic | | Number of measurements |
|---|---|---|
| Californian earthquake, 18 April 1906 | $v = 3.847 \pm 0.045$ km/s | 9 |
| Colombia, 31 Jan. 1906 | $3.806 \pm 0.046$ | 18 |
| Honduras, 1 July 1907 | $3.941 \pm 0.022$ | 20 |
| Nicaragua, 30 Dec. 1907 | $3.916 \pm 0.029$ | 22 |
| **2. Continental** | | |
| California, 18 April 1906 | $v = 3.770 \pm 0.104$ km/s | 5 |
| Philippines I, 18 April 1907 | $3.765 \pm 0.045$ | 30 |
| Philippines II, 18 April 1907 | $3.768 \pm 0.054$ | 27 |
| Bokhara, 21 Oct. 1907 | $3.837 \pm 0.065$ | 19 |
| Bokhara, 27 Oct. 1907 | $3.760 \pm 0.069$ | 11 |

In spite of the fact that individual figures overlap here and there, on the average there is a significant difference in that the velocity of superficial propagation of waves across the ocean floors is about 0.1 km/s higher than that for the continents, and this agrees with the theoretical value expected from the physicial properties of volcanic (plutonic) rock.

Tams has also attempted to combine the observations of as many quakes as possible and to arrive at an average velocity. From velocity values of 38 Pacific quakes he obtained the mean velocity of $3.897 \pm 0.028$ km/s; from 45 Eurasian or American quakes, $3.801 \pm 0.029$. These are the same values as given above.

Another author, Angenheister [47], investigated in 1921 the seismic difference between ocean basins and continental blocks in a number of Pacific quakes; he, too, was concerned with surface waves. He makes a distinction between the two types of wave, transverse and Rayleigh, which were not treated separately by Tams, and thus finds considerably greater discrepancies (though of course his findings are based on only few data): "The velocity of the L waves under the Pacific is 21–26% higher than under the continent of Asia." We should also add that he found characteristic differences for other types of wave: "The transit times for P (*undæ primæ* = primary waves, longitudinal body waves) and S (*undæ secundæ* = secondary waves, shear waves of similar path through the interior of the earth) under the Pacific for 6° focal distance (for distances as short as this these waves are propagated through surface layers only) are 13 and 25 s shorter, respectively, than under the European continent. This corresponds to an 18% increase in S-wave velocity under the ocean.... The period of the posthumous waves is greater under the Pacific than under Asia." All these differences point unanimously in the direction of our theory that the sea floor consists of another, denser material.

Visser also came to the same conclusion regarding surface waves [48]. He found the velocities to be:

over continental regions — 3.70 km/s
over oceanic regions — 3.78 km/s

A velocity differential for surface waves of the same order was also found by Byerly [223] for the Montana earthquake of 28 June 1925.

Finally, Gutenberg corroborated this result by another method [44, 45]. He made use of transverse waves, i.e., surface waves which directly precede the Rayleigh surface waves (and are often not distinguishable from them). The velocity of these waves depends firstly on their wavelength or period, but also on the thickness of the uppermost layer of crust in which they travel. Since not only the transit time (hence velocity) but also the period can be inferred from the seismograms, the thickness of the crustal layer can be found. Admittedly the results of measurement are always rather inaccurate, and a large number of occurrences with different periods are used for the same region in order to draw any conclusion about the layer thickness. Figure 13 gives Gutenberg's results for three regions: (a) Eurasia, (b) for waves that travelled predominantly across the Atlantic floor, (c) the Pacific floor. The abscissa is the period, the ordinate the wave velocity. If the measurements were free of error, all points would have to lie on curves whose position on the figure would depend on the layer thickness. In (a) and (b) three such theoretical curves are drawn for the layer thicknesses 30, 60 and 120 km; in (c) several for zero thickness. Gutenberg concludes that for Eurasia the points fit best the 60-km layer curve, for the predominantly Atlantic-floor area the 30-km layer, but for the Pacific floor the zero-thickness curve. The scatter is large and the method therefore not very precise. However, Gutenberg later found further support for his result. The main point adduced is that in the Pacific the

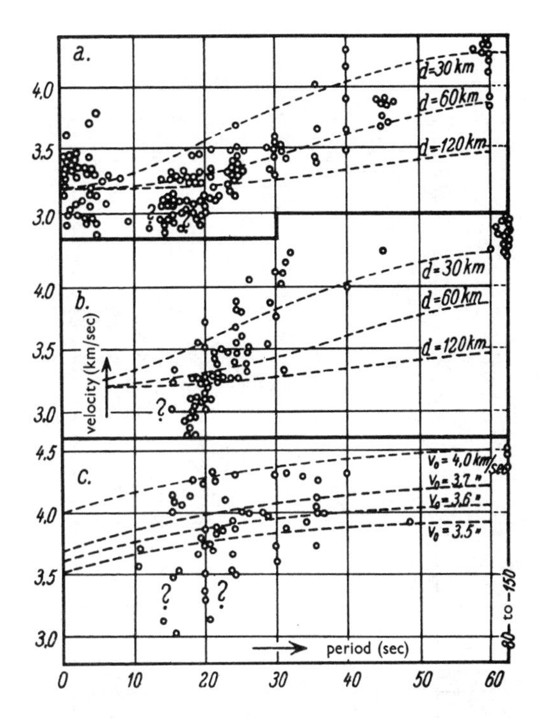


FIG. 13. Velocity of transverse (surface) waves, according to Gutenberg.

topmost layer is apparently absent according to this investigation, as well, and that for wave paths predominantly across the Atlantic, that is, partly via ocean areas and partly via continental areas, the average layer thickness lies between zero and 60 km.[10]

As mentioned before, Angenheister found that the period of posthumous waves, too, is larger in the Pacific region than on the continent of Asia. The question was investigated more closely by Wellmann [49], who corroborated Angenheister's result. Wellmann assembled his data clearly in graphical form (Fig. 14), indicating the foci of the quakes he examined by means of crosses or solid circles, depending on whether the quake emitted long- or short-period posthumous

[10] Gutenberg prefers to regard the result for the Atlantic region as a point against drift theory. In my view this is wrong, and the matter will be brought up again in Chapter 11.

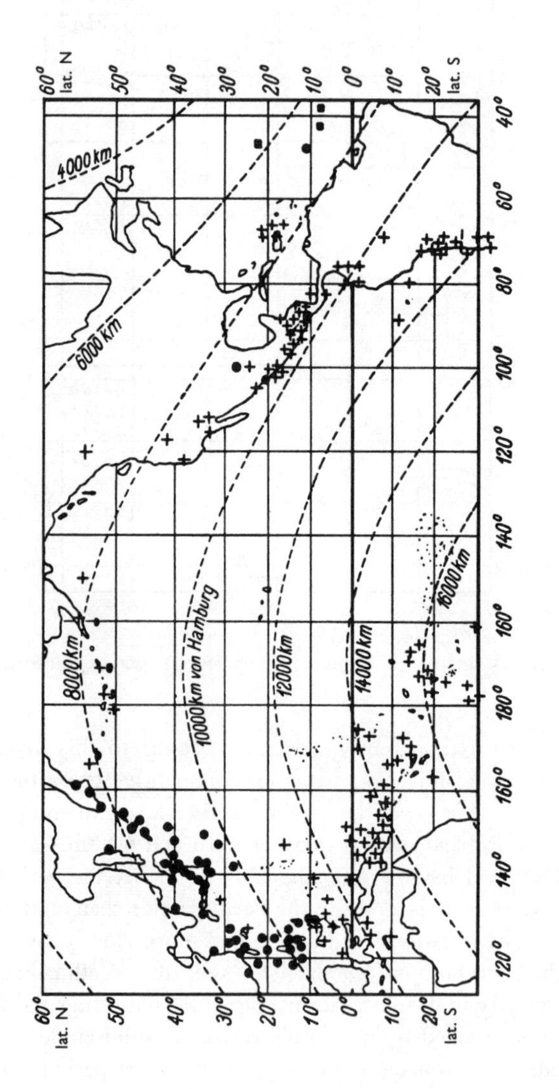


FIG. 14. Earthquake foci, posthumous waves from which were recorded in Hamburg with long (+) or short (●) periods (according to Wellmann).

waves as recorded in Hamburg. If one recalls that the wave path from the focus to Hamburg must lie perpendicular to the dashed lines of equal distance from Hamburg, the figure shows clearly that waves coming from the crosses travelled preferentially over the ocean regions (Pacific, North Sea, North Atlantic), while those emanating from the black circles must have travelled preferentially over continents (Asia).

One can therefore see that seismic research as developed recently has arrived by totally different and mutually independent routes at the conclusion that the ocean floors are of fundamentally different material from that of the continental blocks, and that their material corresponds to a deeper layer of the earth.

As A. Nippoldt brought to my attention, in geomagnetic research the view is generally taken that the ocean floors consist of more strongly magnetisable, and therefore probably more ferriferous material than the continental blocks. This question appears prominently in Henry Wilde's discussion [50] of the magnetic model of the earth, in which the ocean areas have to be covered with sheet iron in order to obtain a magnetic field distribution which corresponds to that of the earth. A. W. Rücker described this experiment in the following way: "Wilde produced a good magnetic model of the earth by means of a setup consisting of a primary field from a sphere magnetised uniformly and a secondary field from lumps of iron laid adjacent to the surface and magnetised by induction. The main bulk of this iron is placed under the oceans.... Wilde considers that coating of the ocean regions with iron is the most important factor." Raclot [52] has also confirmed recently that this experiment of Wilde's is a good representation of the pattern of the earth's magnetic field in its main outlines. Of course, there has been no success so far in calculating the difference between continents and oceans from observations on the earth's magnetic field. This failure is apparently due to a superposed perturbation field of much greater strength whose origin is still unknown and which bears no relation to the distribution of continents and probably cannot do so, as would appear to follow from its large secular variations. In any case, however, the data derived from a study of geomagnetism do not in any way contradict the assumption that deep-sea floors consist of more ferriferous rocks, even in the opinion of such experts as A. Schmidt, who are unwilling to recognise the validity of Wilde's experiment without some reservations. Since, as is well-known, there is general agreement on the fact that even in the silicate mantle of the earth the iron content increases with depth and that furthermore the core consists predominantly of iron, the implication here is that the ocean floor is a deeper layer than that of the continents. At red heat, magnetic effects generally disappear in solids, and if we assume the usual geothermal gradient, magnetic properties would already vanish at depths of about 15–20 km. The strong magnetic field of the ocean floors must therefore be a property of the upper layers, and this agrees well with our idea that the more feebly magnetised material is lacking in such layers.

This strongly suggests the question whether one might not be able to procure test samples of the deep-layer rock from the ocean floor itself. However, it will be impossible for a long time yet to bring up samples of rock outcrop from these depths either by dragnet or other means. Nevertheless, it is worth noting that, according to Krümmel [30], the bulk of the loose samples brought up by dredging is volcanic: "In particular, pumice predominates . . ., then one encounters pieces of broken-off sanidine, plagioclase, hornblende, magnetite, volcanic glass and their decomposition product palagonite, also fragments of lava (basalt, augite-andesite, etc.)." Now, volcanic rock is in fact distinguished by greater density and iron content and is generally considered to have come from deep-lying layers. Suess called this whole group of basic rocks, whose chief representative is basalt, "sima" from the initial pair of letters of each of the main constituents, silicon and magnesium; this was in contrast to the other (silica-rich) group "sal" (from silicon and aluminium), whose chief representatives, gneiss and granite, form the substratum of our continents.[11] Following a suggestion made in a letter written to me by Pfeffer, I prefer to use the term "sial" instead of "sal" to avoid identification with the Latin word for "salt". From the foregoing the reader will probably already have drawn for himself the conclusion that rocks of the sima group, which, to be sure, we know only as volcanic rocks on the sialic continental blocks, where they appear as foreign bodies, have their proper place underneath these blocks and probably also form the ocean floors. It would appear that basalt has the properties required for the sea-floor materials.

In the meanwhile, the question of what materials constitute the

[11] This subdivision goes back to Robert Bunsen, who classified the non-sedimentary rocks into "ordinary trachytic" (silica-rich) and "ordinary pyroxenic" (basic). But it was Suess who coined the convenient names.

different layers of the earth has become the object of many investigations in recent years, partly petrographical and geochemical, partly seismological. The matter is now still in such a state of flux that even partial agreement among the various researchers has not yet been reached. We therefore prefer to be content here with a short survey of the sometimes widely divergent results, without adopting any particular viewpoint of our own.

At first the general point of departure was that it is sufficient to assume a layer of sima up to about 1200 km thick beneath the continental sial layer, which certainly consists of gneiss or granitic material. This sima layer is the mantle. Below this lies the interlayer, down to 2900 km, and then comes the core, consisting essentially of nickel–iron. The interlayer, following the analogy of the sequence of materials in meteorites, may consist of siderolite (pallasite); or, imagining it as the result of a foundry process, pyrites and other ores (slag). It is now a fact, established once and for all, that these are the chief layers of the earth. However, the question whether the sima layer is a single material or whether further subdivision has to be made has been answered in different ways. V. M. Goldschmidt declared that eclogite is the typical representative material of the sima, Williamson and Adams suggested peridotite or pyroxenite, others dunite. In any case, the bulk of the sima must be a very basic or "ultra-basic" rock, more basic than basalt, so that the latter is most probably the topmost layer of the sima. A large number of papers and some books by Jeffreys [53], Daly [54], S. Mohorovičić [55], Joly [56], Holmes [57], Poole [58], Gutenberg [59], Nansen [222] and others have considered the questions which arise here. It is specially noteworthy that Daly's book (*Our Mobile Earth*, London, 1926) is based altogether on the drift theory; Joly's book (*The Surface History of the Earth*, Oxford, 1925) attacks the theory but does in fact adduce important new evidence in its favour by considering radioactive heat generation.

All writers apparently agree that basalt is the first material to be encountered below the granite of the continental blocks. However, the boundary layer between these two materials is no longer identified by most research workers with the large layer at 60 km deduced from seismology; it is assumed to lie at 30–40-km depth, where earthquakes also enable one to detect another boundary layer, although of less importance. One of the chief reasons why granite is not readily assumed to extend down to 60 km is that a layer of this thickness would contain too much radium and would therefore produce too much heat. The ultra-basic material (dunite, etc.) would then start at 60 km. Furthermore, Mohorovičić in particular has laid emphasis on the fact that the 60-km boundary interlayer shows no variation in radial position under mountains and plains, whereas the boundary between granite and basalt, which lies nearer the outer surface, in fact does so. The question therefore arises whether in these circumstances one should not rather regard the granite layer at 30–40-km depth as the lower limit of the continental blocks, instead of the large layer at 60 km as heretofore. On the other hand, there is still no explanation of how the latter boundary layer behaves under the oceans. Gutenberg assumes that the large boundary layer at 60-km depth forms the sub-Pacific surface, so that here the ultra-basic material (dunite) would also appear, as outcrop. Mohorovičić, however, believes that the ocean floors are made of basalt.

We shall have to await further developments in these investigations before it will be possible to construct a final picture. However, it may very well be possible that this multiplication of layers will also introduce complications so far as the nature of the ocean floors is concerned; signs of this have already appeared in another connection (p. 39).

Nevertheless, regardless of how the various viewpoints may develop, this much is already clear: they are progressing along the same lines as drift theory, because the fundamental contrast between ocean floors and continents is no longer denied, and for drift theory it is all one whether the former are basalt or perhaps consist here and there of ultra-basic material. In any case, apart from residue, the granite cover of the continental blocks is lacking on the ocean floors.

A frequent objection to drift theory is this: The earth is as solid as steel and therefore the continents cannot move. Actually, studies of earthquakes, polar fluctuations and the tidal deformations of the solid earth have all led to the same result: the coefficient of plasticity or the stiffness coefficient of the earth averages out at $2.10^{12}$ g/cm$\cdot$s$^2$; or, if one distinguishes between a rock mantle extending to 1200 km depth and an ore/metal core, the value for the former is $7 \cdot 10^{11}$ and for the latter $3 \cdot 10^{12}$. Since this coefficient for cold steel is $8 \cdot 10^{11}$, the earth is in fact as stiff a material as steel. But what follows from this? In the first instance, nothing, because the velocity at which a continent can move under the influence of a given force depends, after all, not on the plasticity (stiffness) of the sima but on another, independent property of the material, "internal friction" or "viscosity," or on its reciprocal, "fluidity." Viscosity has the dimension g/cm·s. Unfortunately the viscosity cannot reliably be deduced from the plasticity, but has to be determined by special experiments. Measurements of viscosity on so-called solids are extremely difficult. Even in the laboratory, where use is made of damping of elastic vibrations, rate of deformation in bending or torsion, or measurement of so-called relaxation times, such measurements have only been carried out on a very few substances. At the moment, unfortunately, it is an almost hopeless task to clear up the question of the viscosity coefficient of the earth. To be sure, there have been recent attempts of various kinds to estimate the viscosity coefficient of the earth, partly as an overall average, partly for certain layers, but the results are so discrepant that all we can establish is our complete ignorance of the matter.

All that can be said with certainty is that the earth behaves as a solid, elastic body when acted upon by short-period forces such as seismic waves, and there is no question of plastic flow here. However, under forces applied over geological time scales, the earth must behave as a fluid; for example, this is shown by the fact that its oblateness corresponds exactly to its period of rotation. But the critical point in time where elastic deformations merge into flow phenomena depends precisely on the viscosity coefficient.

In his investigation of the detachment of the moon from the earth, G. H. Darwin assumed that tidal forces acting for 12 and 24 hours give rise to flow deformation, and this hypothesis has been applied by many other writers. In a more recent investigation, however, Prey [60] came to the conclusion that Darwin's assumptions do not imply that perhaps even today the earth's crust is being displaced appreciably westwards by the tidal friction. Fifty to sixty million years ago the earth's viscosity coefficient may still have had the relatively low value of about $10^{13}$ (roughly that of glacier ice), and at that time, according to Prey, large crustal displacements therefore occurred. Since then, however, he says, the viscosity coefficient must have risen to the point where such displacements are now impossible. One should note here, of course, that Darwin was not yet in a position to take the radium content of the crust into consideration. Prey assumes a progressive cooling process in spite of the radium. All the same, our present-day knowledge of the amount of radium present and of the geological facts would lead us to doubt very much whether in the course of the geological epochs, estimated to be enormously long, the earth's viscosity coefficient has appreciably altered in a systematic way, apart from fluctuations.

Geologists have often assumed that there is a magma layer under the solid crust of the earth, and in the same way Wiechert believed that certain peculiarities of seismograms might be explained by such a fairly fluid layer. Schweydar [61] objected to this on the grounds of the measurable tidal movements of the solid earth. If, in fact, fluidity contributed perceptibly to these movements, they would lag behind the periodicities of the sun and moon. Since, however, no such time lags are observed, the amount of tidal movement must be a function of elasticity only, not of plasticity, or fluidity. The margin of error in the observations thus provides at the very least a limiting value for the viscosity coefficient, which of course turns out to be different according to the thickness of the layer to which it is assumed to apply. This is because a low-viscosity, thin layer gives the same displacements as a high-viscosity layer of correspondingly greater thickness. Schweydar thus finds that the viscosity coefficient must be more than $10^9$ when the layer in question is only 100 km thick, but more than $10^{13}$ or $10^{14}$ if this layer is 600 km thick. Naturally it is a basic assumption here that the layer is a coherent one which covers the whole earth. Small-area, isolated portions of the earth may be considerably more plastic.

Schweydar made another attempt in 1919 to determine the viscosity of the earth in his investigation of polar wandering [62]. He calculated backwards to find out how large the polar fluctuation periods would have to be if half the viscosity coefficient of the earth had the values $10^{11}$, $10^{14}$, $10^{16}$ and $10^{18}$. He found that in the case of the first two figures the period of polar fluctuation could only be one of about 80 years. Only for the greater viscosities did a short period of 470–370 days result, that is, one of the order of magnitude which actually obtains. Here again it is obviously a question of how thick one assumes the viscous layer to be. If the whole earth is assumed to have the same viscosity, a short period only arises for the viscosity value $10^{18}$, but does so for the value $10^{13}$ if only the layer between 120 and 600-km depth is assumed to be plastic. Since it was possible to carry out the computation only for constant density throughout the earth, the result can be regarded as but a first approximation to fix ideas. Subsequently, Schweydar used the viscosity $10^{19}$ under the assumption that only the layer between 100 and 1600-km depth is fluid.

Schweydar advocates high values of the earth's viscosity. For all that, he himself concludes: "Nevertheless, one must admit the possibility that the continents may experience a displacement towards the equator under the influence of the force exerted away from the poles" [40]. Later on we shall discuss the essential facts about the contra-polar driving force and the calculation which leads to this result.

Jeffreys [53] has assumed still higher viscosities, namely $10^{21}$ in the layer where the value is least. So far as I know, this is the most extreme view of any.

On the other hand, some of the most recent opinions incline to really astonishingly low viscosities, although only in a relatively thin layer. For example, Meyermann [64, 65] starts out from the fact, recently discovered by astronomical means, that the earth's rotation is non-uniform: "In 1700, for example, each point of the earth's surface was about 15 s east, in 1800 about 15 s west, in 1900 about 10 s east and in 1924 about 20 s west of the position it would have occupied if the earth rotated uniformly. Since it is out of the question that the earth as a whole experiences such fluctuations, I regard this as an indication that the crust drifts westwards relative to the core. . . . If the friction increases the drift is less. . . . If it decreases, then conversely the earth's surface moves westwards relative to the hypothetical earth." Both in the components of the earth's magnetism and in the fluctuation of the length of the day, according to Meyermann, there is a periodicity of 270 years; he deduces from this a complete cycle of the crust in the amazingly short time of 270 years, and concludes accordingly that, if fluidity is confined to a zone 10 km thick, the viscosity (internal friction) coefficient is only about $10^3$ in this layer; this is 21 times more viscous than glycerine at 0° C. However, for the time being it must remain undecided whether his interpretation is really in keeping with the facts. In this respect a paper of Schuler's [66] is worthy of note; he shows that when the polar inland icecaps are enlarged, the movement of mass towards the axis of rotation caused by this must, according to the law of conservation of angular momentum, produce a marked acceleration of the earth's spin. Conversely, deceleration must occur when the ice melts and mass transport takes place in the direction of the equator, i.e. away from the axis.

The problem of the viscosity of the layers lying below the continental blocks is closely dependent on whether the temperature of these layers exceeds the melting point or not. Although it is probable that the molten magma can have a very high viscosity at very high pressures and therefore behave as a solid—the phenomena at such pressures are still unknown—yet all authors who support the idea of a molten layer tend towards the view that the viscosity of this layer is low enough to permit large displacements, convection currents in fact. Consideration of the radium content has produced quite new viewpoints on precisely this question.

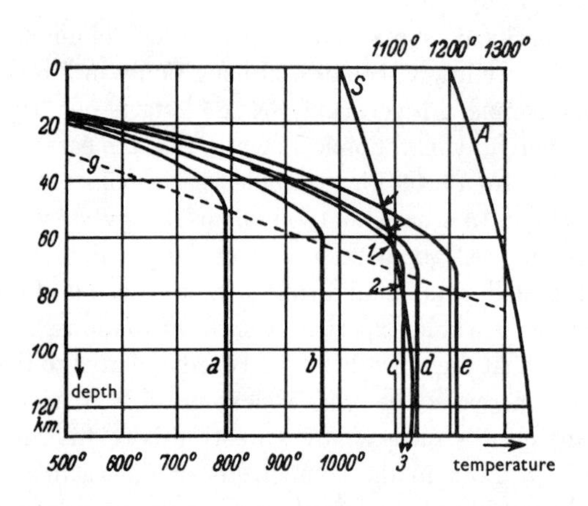


FIG. 15. Depth-dependence of temperature (*a* to *e*) and fusion temperatures (*S* and *A*) down to 120 km (according to von Wolff).

Figure 15 gives von Wolff's representation of the temperature distribution in the outer 120 km of the crust. Curves *a* to *e* are calculated with different assumptions about the crustal radium content. Additionally, two melting-point curves *S* and *A* are plotted. Here, too, different curves are obtained according to the material assumed to be present. *S* corresponds to the lowest conceivable fusion temperatures for the various depths. As shown by the knee of the temperature curves and the slope of the melting-point curves, there is an optimum region for melting at about 60–100 km down, and it is thus possible that here a molten layer is confined between two crystalline layers.

It is natural to ask whether seismology might not provide an answer to the problem. Unfortunately, this is not so; it could do so if the molten state implied a low viscosity or fluidity, since in a fluid medium there can be no propagation of transverse waves such as the *S* waves. It is generally believed nowadays, however, that any such material heated above the melting point and therefore molten or fused exists in an amorphous and glassy (therefore solid) state. Nevertheless, seismology does give a pointer here. It can be shown that, making the most likely assumptions about the density of the material, its elastic resistance to deformation, which otherwise generally increases with depth, exhibits a discontinuity in this behaviour at about 70-km depth and may even show a transitory reduction in resistance. Writers such as Gutenberg [104] explain this by saying that, in all probability, at these depths the crystalline state is converted to the amorphous, glassy one. Even if the glassy state should be considered as solid where short-period seismic waves are involved, it is nevertheless quite likely that it exhibits an appreciable degree of fluidity under forces which operate over geological time scales.

There are also certain geological facts to be considered here. The rare and large-scale "granite fusions" of southern Africa described by Cloos [103] show that, in some periods of the earth's history, the fusion isotherm of granite has been pushed locally to just below the surface. This is all the more reason to believe that at such times the rock at depths of 60–100 km must have been molten. The isothermal surfaces in the earth have quite certainly no fixed positions, but vary in both time and space. Joly [56] explains this by saying that below the continental blocks excess heat is generated by radioactivity and that therefore the temperature is continuously rising; this reaches the point where melting occurs and the blocks are floated. They then move over cooler portions of the globe which were formerly ocean regions. A fact which strongly supports this idea is that the geothermal interval (per degree rise) in Europe averages 31.7 m and in North America 41.8 m. This marked difference, which has been much discussed recently, implies that the interior of the earth under North America is cooler than under Europe. Daly is probably right when he states: "A conceivable explanation may be found in the comparatively recent sliding of North America over the sunken crust of the old, Greater-Pacific basin" [67].

At this point we should mention the authors who attribute the phenomena of the outermost crust to "undercurrents," such as Ampferer [68] and Schwinner [69], among others. According to Ampferer, undercurrents have dragged America westwards; Schwinner believes that there are convection currents in the liquid layer caused by non-uniform output of heat, and that these currents draw the crust along and compress it at areas where they take a downward path. In connection with the excess radioactive heat production in continental blocks, Kirsch [70] has made extensive use of the idea of such thermal convection currents in the fluid layer. He assumes that the continents were at one time joined together and that excess heat was generated beneath them (recall the southern African granite fusions); this led to circulation of the fluid substratum, which flowed outward everywhere towards the ocean basins, and there moved downwards due to increased thermal loss, while rising under the centre of the continental region. The continental platform was finally broken up by the friction and the fragments were separated in all directions by the current. Kirsch arrived at astonishingly high flow rates for the current and correspondingly low viscosity values in the molten layer.

All these approaches show one thing: we should not be dogmatic about the viscosity coefficient of the earth's interior and in particular that of the individual layers, because we still know nothing about it. Schweydar's results are fundamentally inconclusive because they do not exclude the possibility of a *discontinuous*, relatively fluid layer, and obviously have nothing to say on the question of whether there can have been such a relatively fluid continuous layer during certain periods of prehistory. His results are, however, of great value in that they lead to viscosity values which permit continental drift even though the idea of a fluid layer is rejected. The possibility of drift therefore does not depend on the ultimate correctness of those authors who have recently championed the existence of a fluid layer beneath the continental blocks, at least in certain regions and at certain periods.

After the foregoing, it is superfluous to state that the drift theory agrees excellently with the results of geophysics. Indeed, it forms a starting point for a large number of promising new lines of research, which have already yielded important data, even though many details will be completely explained only in the future.

One could adduce many other observational facts in geophysics which would lend support to drift theory, either directly or indirectly, but it is not possible within the scope of this book to deal comprehensively with the different subjects relevant to the problem, or even to attempt to do so. Some of the facts will be discussed in later chapters.

CHAPTER 5

# *Geological Arguments*

BY COMPARING the geological structure of both sides of the Atlantic, we can provide a very clear-cut test of our theory that this ocean region is an enormously widened rift whose edges were once directly connected, or so nearly as makes no difference. This is because one would expect that many folds and other formations that arose before the split occurred would conform on both sides, and in fact their terminal sections on either side of the ocean must have been so situated that they appear as direct continuations of each other in a reconstruction of the original state of affairs. Since the reconstruction itself is necessarily unambiguous because of the well-marked outlines of the continental margins and allows no scope for juggling, we have here a totally independent criterion of the highest importance for assessing the correctness of drift theory.

The Atlantic rift is widest in the south, where it first started. The width here is 6220 km. Between Cape São Roque and the Cameroons there is a gap of only 4880 km, between the Newfoundland Bank and the shelf of Great Britain only 2410 km, only 1300 km between Scoresby Sound and Hammerfest and probably only 200–300 km between the shelf margins of northeastern Greenland and Spitsbergen, where the rift appears to have occurred only in relatively recent times.

Let us begin by comparing the southern margins. In the southernmost part of Africa there is a Permian folded range striking east–west, the Swartberg. In the reconstruction, the westerly extension of this range meets the area south of Buenos Aires which, according to the map, does not seem to be marked by any special feature. It is very interesting to note that Keidel [72, 73] found in the local sierras, particularly the more strongly folded southern portion, ancient folds which completely resembled in structure, rock series and fossil content not only the pre-cordilleras of San Juan and Mendoza Provinces to the northwest, which abut the Andean folds, but what is more, the South African Cape mountains. He states: "In the sierras of Buenos Aires Province, particularly in the southern range, we find a succession of beds very like that of the Cape mountains of South Africa. There appears to be strong conformity among at least three cases: the lower sandstone of the Lower Devonian transgression, the fossil-bearing schists which mark the culmination of this transgression and a more recent and very characteristic structure, the glacial conglomerate of the Upper Palæozoic. . . . Both the sedimentary rocks of the Devonian transgression and the glacial conglomerate are strongly folded just as in the Cape mountains; and here, as there, the direction of the folding movement is mainly a northerly one." All this is an indication that we have here an elongated, ancient fold that traverses the southern tip of Africa, then is continued across South America south of Buenos Aires and finally turns north to join the Andes. Today the fragments of this fold are separated from each other by an ocean more than 6000 km wide. In our reconstruction, which here in particular permits of no manipulation, the fragments are brought into direct contact; their distances from Cape São Roque in one case and the Cameroons in the other are equal. This evidence for the correctness of our synthesis is very remarkable, and one is reminded of the torn visiting card used as a means of recognition. The fact that at the African coast the Cedar Berg chain branches north from the main run of the South African range has little bearing on the transatlantic conformation. This branch soon disappears and has the appearance of a local deflection which might have been produced by some discontinuity at the point of subsequent rifting. We can see such branches as this much more frequently in the case of the European fold mountains of both the Carboniferous and the Tertiary and there is nothing here, either, to prevent us regarding these folds as a single system originating from a single cause. Even though recent investigations have made it appear that the African fold system has lasted down to recent times, this does not imply that there is a difference in age, for Keidel states: "In the sierras, the glacial conglomerate, the most recent formation, is folded; in the Cape mountains, the Ecca beds at the base of the Gondwana series (Karroo strata) still show signs of the movements. . . . In both regions, therefore, the main movements could have occurred in the period between the Permian and the Lower Cretaceous."

However, this corroboration of our viewpoint provided by the Cape mountains and their continuation in the sierras of Buenos Aires is by no means the only one; we can find many other items of evidence along the Atlantic coastlines. Even in its broad outlines the vast gneiss plateau of Africa, last folded a long time ago, shows a striking similarity to that of Brazil, and this similarity is not confined only to generalities, as is revealed by the conformity between the igneous rocks and between the sedimentary deposits of each area, and by the conformity between the original fold directions.

H. A. Brouwer [74] has made a comparison of the igneous rocks. He finds no less than five parallels: (1) the older granite, (2) the younger granite, (3) alkali-rich rocks, (4) volcanic Jurassic rock and intrusive dolerite and (5) kimberlite, alnoite, etc.

In Brazil, the older granite is found in the so-called "Brazilian complex"; in Africa, in the "fundamental complex" of the southwest, and also in the "Malmesbury system" of southern Cape Colony and in the "Swaziland system" of the Transvaal and Rhodesia. Brouwer says: "Both the east coast of Brazil in the Serra do Mar and the opposite west coast of southern and central Africa consist mainly of these rocks, and in many ways they give the landscape of both continents a similar topography."

The more recent granite is, on the Brazilian side, intrusive in the "Minas series" of the states of Minas Geraes and Goyaz, where it forms gold-bearing lodes, and is also an intrusive rock in São Paulo State. In Africa, the corresponding rock is the Erongo granite of Hereroland and the Brandberg granite of northwestern Damaraland as well as the granites of the "Bushveld igneous complex" in the Transvaal.

The alkali-rich rock is also to be found on exactly corresponding stretches of coastline: on the Brazilian side, at various places in the Serra do Mar (Itatiaya, Serra do Gericino near Rio de Janeiro, Serra de Tingua, Cabo Frio); on the African side, on the coast of Lüderitzland, at Cape Cross north of Swakopmund, but also in Angola. Farther from the coastline, there are the associated volcanic areas about 30 km in diameter, Poços de Caldas in the south of Minas Geraes State and Pilandsberg in the Rustenburg district of the Transvaal. These alkaline rocks, especially, are very striking in their completely similar plutonic, gangue and extrusive rock formations.

Referring to the fourth group of rocks (Jurassic volcanic rocks and intrusive dolerite), Brouwer says: "Just as in South Africa, there is a thick series of volcanic rocks in the bottom section of the Santa Catharina system, which corresponds approximately with the South African Karroo system; the series can be considered as Jurassic and covers large areas in the States of Rio Grande do Sul, Santa Catharina, Parana, São Paulo and Matto Grosso, and even in Argentina, Uruguay and Paraguay." In Africa we have the Kaoko formation between latitudes 18 and 21° S, where a similar type of rock corresponds to that in the southern Brazilian States of Santa Catharina and Rio Grande do Sul.

Finally, the last of the rock groups (kimberlite, alnoite, etc.) is the best-known, since both in Brazil and South Africa the beds yield the famous diamond finds. In both these regions the peculiar type of stratification known as "pipes" occurs. There are white diamonds in Brazil in Minas Geraes State and in South Africa north of the Orange River only. However, the correspondence between the two regions is shown more clearly by the extent of the kimberlitic parent rock than by these rare diamond sites. The same thing has been established in the gangues of Rio de Janeiro State: "As in the case of the kimberlite rocks near the west coast of South Africa, the well-known Brazilian rocks almost all belong to the low-mica basaltic varieties."[12]

Brouwer stresses, however, that even the sedimentary rocks correspond closely on both sides: "The similarity between some groups of sedimentary rocks on both sides of the Atlantic Ocean is also very striking. We mention only the South African Karroo system and the Brazilian Santa Catharina system. The Orleans conglomerate in Santa Catharina and Rio Grande do Sul matches the Dwyka conglomerate of South Africa, and in both continents the uppermost sections are formed by the thick volcanic rock series already mentioned, like those of the Drakenberg in Cape Colony and of the Serra Geral in Rio Grande do Sul."

Du Toit [75] even conjectured that the erratic Permo-Carboniferous material of South America partly derived from Africa: "The Southern Brazilian tillite was, according to Coleman, derived from a

[12] H. S. Washington also acknowledges these correspondences between the volcanic rocks, but nevertheless he says that a comparison does not favour the drift theory, mainly because he demands too much of the comparison. His rejection of the idea is poorly founded, but unfortunately has had a decisive effect on the attitudes of many American geologists.

sheet probably having its centre to the southeast,[13] off the present coast line. Both he and Woodworth also record certain erratics of a peculiar quartzite or grit with banded jasper pebbles, which from their accounts are just like those collected by the Transvaal ice from the ranges of Matsap beds in Griqualand West and transported so far westwards at least as the 18th meridian. With the continental disruption hypothesis in mind, could they not possibly have been carried much farther westwards still?" Recently, however, L. C. Ferraz (cited in [78]) found this rock as an outcrop south of the place of its discovery near Blumenau in Santa Catharina, on the north bank of the Itajahý river; the interpretation suggested by du Toit therefore loses its force. On the other hand, the similar occurrence of the outcrop in both Brazil and South Africa is another very noteworthy link in the long chain of striking conformities between the two continents.

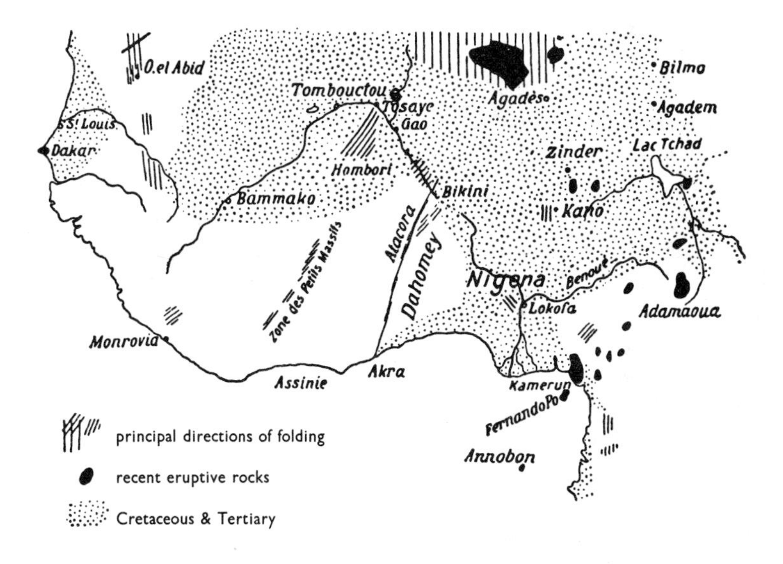


Fig. 16. Strike directions in Africa (according to Lemoine).

We find additional conformities in the directions of the ancient folds which extend throughout these large gneiss plateaux. In the case of Africa, we refer to the map drawn by Lemoine [76] are shown in Figure 16. It was drawn up for other purposes and therefore does not

[13] The original text has "south-west", which is clearly a slip, as the words which follow it prove.

illustrate what we require very clearly, but nevertheless does show it. In the gneiss massif of the African continent there are two main strike directions (trend lines) of somewhat different age. In the Sudan the predominant one is the older, northeasterly strike, which is at once evident in the straight upper course of the Niger, running in a similar direction and visible as far as the Cameroons. It cuts the coastline at 45°. However, south of the Cameroons and still just observable on the map, the other, younger strike direction takes precedence, running roughly north–south and parallel to the curves of the coast.

In Brazil we find the same phenomenon. E. Suess wrote: "The map of eastern Guiana . . . shows the more or less east–west strikes of the old types of rock that constitute this area. The included Palæozoic strata which form the northern part of the Amazon basin also follow this direction, and the run of the coast from Cayenne towards the mouth of the Amazon is therefore across the strike direction. . . . So far as the geology of Brazil is known today, it forces one to assume that up to Cape São Roque the contour of the mainland crosses the strike direction of the mountains, but that from these foothills on, all the way down to Uruguay, the lie of the coast is marked by the mountains." Here also the courses of the rivers (the Amazon on one side, the Rio São Francisco and the Parana on the other) generally follow the strike direction. Of course—and this is shown in the tectonic map of South America drawn by Keidel (*loc. cit.*) which essentially follows J. W. Evans (Fig. 17)—recent investigations prove that there is a third strike direction parallel to the northeast coast; the situation thereby becomes somewhat more complex. However, the other two strike directions show up very clearly in this map, although somewhat off the line of the coast in some parts. Taking into account the large angle through which South America must be turned in our reconstruction, the direction of the Amazon becomes exactly parallel to that of the upper course of the Niger, so that the two strike directions coincide with the African ones. We may see in this a further confirmation for the direct connection that once existed between the two continents.

The similar structure of Brazil and southern Africa has been emphasised more and more strongly of late. Maack [77] states: "Anyone who knows southern Africa will find the geology of this (the Brazilian) landscape startling. At every step I was reminded of the formations of Namaland and the Transvaal. The Brazilian strata correspond

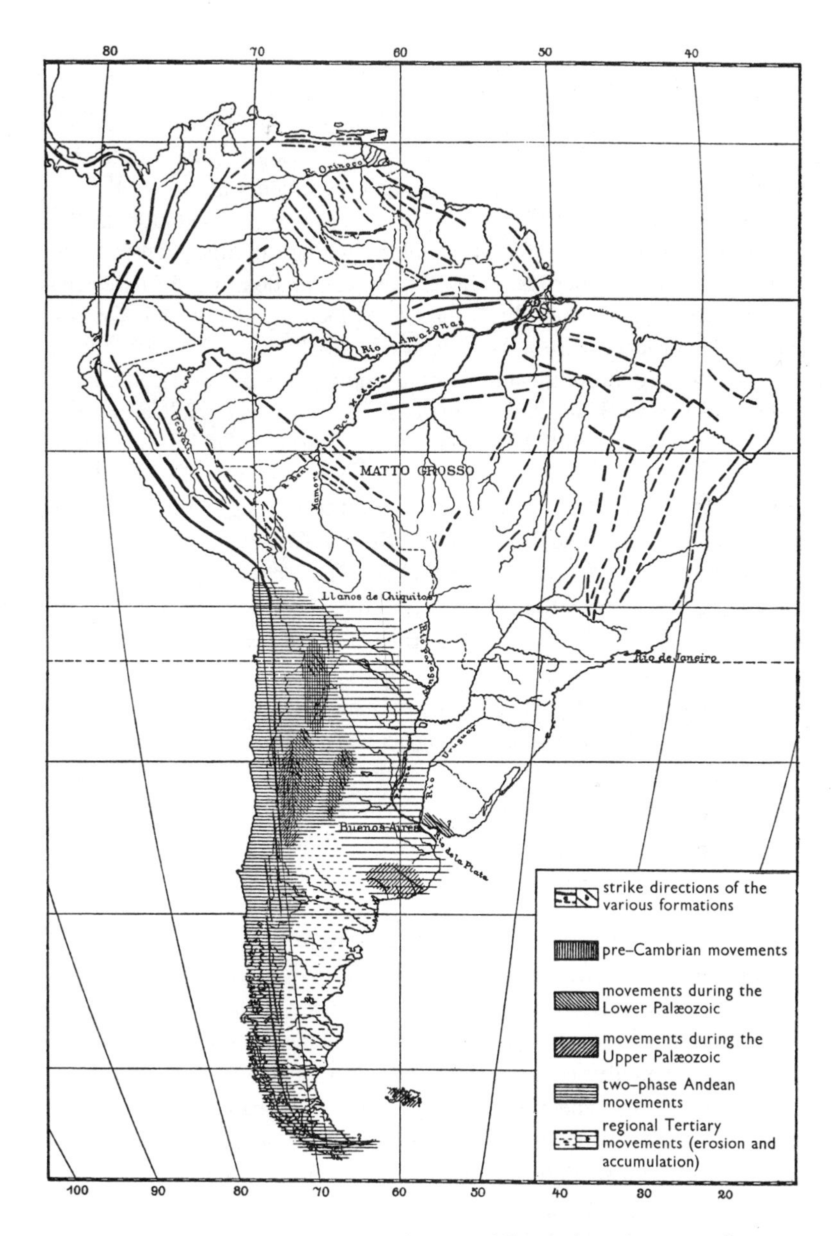


FIG. 17. Diagrammatic tectonic map of South America, according to Keidel and J. W. Evans.

perfectly in every detail to the strata series of the southern African shield." On this journey Maack found five kimberlite pipes at Patos (ca. 18½° S, 46½° W). He concludes: "It is obvious that in view of the distance which today separates the corresponding formations, one must reject the idea of sunken land bridges which extended across the Atlantic. A displacement of continents in A. Wegener's sense is what comes to mind, a concept which finds support in the observation that, since the very oldest geological times, apart from the Permo-Carboniferous, a dry climate has predominated in western South Africa, and that the Triassic sedimentary deposits in Minas are in accordance with a dry inland climate."

Particularly thorough comparative studies have been carried out by the well-known South African geologist du Toit, who made a journey of exploration in South America for this purpose. The results of this investigation, which includes a very complete survey of the literature, were published in 1927 as No. 381 of the Carnegie Institution of Washington (157 pages) with the title *A Geological Comparison of South America with South Africa* [78]. The whole work is a unique geological demonstration of the correctness of drift theory so far as these parts of the globe are concerned. If we wanted to cite every detail in the book which favours the theory, we would have to translate it from start to finish. There are many statements like the following: "Indeed, viewed even at short range, I had great difficulty in realising that this was another continent and not some portion of one of the southern districts in the Cape . . ." (p. 26). On page 97 he says: "In preparing this review an attempt was made first of all to write the historical account, irrespective of any hypothesis as to the manner of such union or of the ultimate mode of separation of the land-masses though it became evident, as the data were assembled, that they pointed very definitely in the direction of the displacement hypothesis. . . ." Conformities between the two sides of the ocean, he says, are now known in such numbers that it is no longer possible to imagine them accidentally coexisting, particularly since they cover vast stretches of land and the time span from the pre-Devonian to the Tertiary. Du Toit adds: "Furthermore, these so called 'coincidences' are of a stratigraphical, lithological, palæontological, tectonic, volcanic and climatic nature."

We cannot reproduce here even the summary of points of conformity which fill seven pages of du Toit's Chapter VII ("Bearing on the

Displacement Hypothesis"). However, we give the comparison of the chief geological features assembled on pages 15 and 16:

"Confining attention in each case to a strip some 45 degrees in length by 10 degrees in breadth, we shall now proceed to compare the two stretches, namely, the tract extending from Sierra Leone to Cape Town on the one side with that from Pará to Bahía Blanca on the other. . . .

"In *each continent*:

"(1) The foundation rocks consist of crystallines of pre-Cambrian age and certain belts of infolded pre-Devonian sediments of various though mostly undetermined ages, but generally corresponding lithological characters.

"(2) In the extreme north, marine Silurian and Devonian beds, only slightly disturbed, rest uncomformably upon this complex, occupying a broad syncline trending obliquely to the coast line, namely, between Sierra Leone and the Gold Coast and underlying the estuary of the Amazon.

"(3) Farther to the south, belts of Proterozoic and Early Palæozoic strata, mainly quartzites, slates and limestones, strike nearly parallel to the coast, being gently flexed in the north, but becoming more disturbed toward the south, where they are invaded by granitic masses, for example, in the region between Lüderitz and Cape Town and that between Río São Francisco and Río La Plata.

"(4) Corresponding to the nearly flat-lying Devonian of Clanwilliam is its all but identical counterpart in Paraná and Matto Grosso.

"(5) More to the south we find the Devono-Carboniferous of the southern Cape paralleled by the terrain appearing a little to the north of Bahía Blanca, passing up conformably into the Carboniferous glacials and Permian sediments, both successions having been intensely crumpled under Permo-Triassic and Cretaceous movements that display similar orientations.

"(6) Traced northwards, the tillites in each case become horizontal and transgress across the Devonian to rest upon a glaciated peneplain formed by these and by older rocks; farther to the north they fail.

"(7) The glacials are in each case overlain by continental Permian and Triassic strata with the "*Glossopteris* flora," covering enormous areas, followed by vast outpourings of basalts and penetrated on an extensive scale by dolerites of presumed Liassic age.

"(8) These Gondwana beds extend northward from the southern Karroo to the Kaokoveld and from Uruguay to Minas Geraes.

"(9) Further great detached areas occur in the north, in each case some distance inland, in the Angola–Congo and the Piauhý–Maranhão regions.

"(10) An intraformational break is widespread, though commonly there is no angular unconformity between late Triassic and early Permian beds. In certain areas, however, the former may rest with visible discordance on tilted Permians or Pre-Permian formations.

"(11) Tilted Cretaceous beds occur on the coast only in the Benguella–Lower Congo and Bahia–Sergipe areas.

"(12) Horizontal Cretaceo–Tertiary, both marine and continental, cover great extents between Cameroons and Togoland and in Ceará, Maranhão, and also more to the south, while the extensive deposits of the Kalahari can roughly be paralleled with the Neogene and the Quaternary Pampean of Argentina.

"(13) In setting down this generalised summary, the important link constituted by the Falkland Islands must not be overlooked. Their folded Devono–Carboniferous succession is all but indistinguishable from that of the Cape, while the Lafonian closely parallels the Karroo system. *Stratigraphically and structurally, the Falklands have their place with the southwest of the Cape and not with Patagonia*. . . .

"(14) From the palæontological viewpoint, attention should be focussed on: (*a*) the 'austral facies' of the Devonian of the Cape, Falklands, Argentina, Bolivia and Southern Brazil, in contrast to the 'boreal facies' of northern Brazil and central Sahara; (*b*) that unique reptilian genus *Mesosaurus* from the Dwyka shales of the Cape and the Iratý shales of Brazil, Uruguay and Paraguay; (*c*) the *Gangamopteris–Glossopteris* flora, with a small admixture of northern forms within the Lower Gondwana beds in the south of each country; (*d*) the *Thinnfeldia* flora of the Upper Gondwana of the Cape and Argentina; (*e*) the Neocomian (Uitenhage) fauna in the south of the Cape and northwest of Neuquén in Argentina; (*f*) the Northern or Mediterranean facies of the Cretaceous and Tertiary faunas north of the Tropic of Capricorn; and (*g*) the South Atlantic–Antarctic facies of the Eocene (San Jorge formation) of Patagonia.

"(15) The geographical outlines of Africa and South America are amazingly similar, not only in the main but even as to detail; moreover, excepting in the north, the fringe of Tertiaries is of small width and the presence of those beds of little moment therefore."

Of special interest here is a quite new factor in the geological relationships between the two continents, one which du Toit was the first to bring out. On page 109 he says:

"*Of prime importance, moreover, is that evidence obtainable from the study of the phasal variations displayed by particular formations when traced within their respective continents.*

"In illustration, let us consider the case of two equivalent formations, the one in South America beginning on or near the Atlantic coast at $A$ and extending westward to $A'$ and the other in Africa starting similarly near the coast at $B$ and stretching eastward to $B'$. Then it can be affirmed that more than one such instance can be designated, where the change of facies in the distance $AA'$ or $BB'$ is *greater* than that found in $AB$, although the full width of the Atlantic intervenes between $A$ and $B$. In other words, these particular formations along the two opposed shores tend to resemble one another more closely than either one or both of their actual and visible extensions within the respective continents. With the multiplication of such examples, drawn from more than one geological epoch, such a singular relationship can no longer be regarded as wholly fortuitous and a definite explanation therefore has accordingly to be sought. An analysis, moreover, shows that this unexpected tendency is equally marked, whether the formations involved be marine, deltaic, continental, glacial, eolian or volcanic."

Du Toit's book contains the chart we reproduce as Figure 18, which gives the relative positions of the two continents before their separation. Du Toit stresses that in this reconstruction one must still leave a gap of at least 400–800 km between the present-day coastlines if it is desired to allow for the difference in facies to be observed there. I am bound to agree completely with this point, for not only must there remain room between the two coastlines for the shelf that extends in front of them, but possibly even for the material composing the mid-Atlantic ridge. The relative position of the blocks may be more accurately fixed perhaps when the numerous echo soundings of the "Meteor" expedition are evaluated and studied. It is my belief that the result will be a picture similar to that presented by du Toit and based on geological comparisons.

Du Toit correctly held that a special corroboration of drift theory is to be found in the fact that the Falkland Islands, although supported on

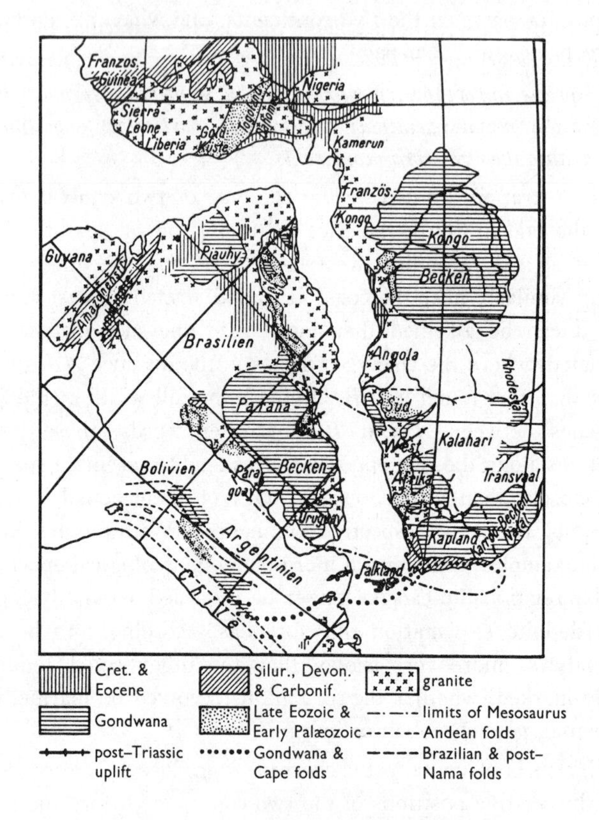


FIG. 18. Former relative position of South America and Africa, according to du Toit.

the coastal shelf off Patagonia, have no geological connection with that area but rather with southern Africa.[14]

I must admit that du Toit's book made an extraordinary impression on me, since up till then I had hardly dared to expect so close a geological correspondence between the two continents.

As I have already shown, on palæontological and biological grounds

[14] I must confess that the position of the Falklands assumed by du Toit (Fig. 18) in his reconstruction seems questionable to me, considering their present-day position and the depth chart of the South Atlantic. I would have placed them south rather than west of the Cape of Good Hope in this reconstruction; however, this is a secondary problem which further research will no doubt elucidate.

it must be concluded that the interchange of forms between the land regions of South America and Africa ceased during the Lower to Middle Cretaceous. This in no way contradicts Passarge's view [79] that the marginal rifts of southern Africa were already formed in the Jurassic period, because the rift opened up only gradually from the south, and above all since the formation of trough faults probably occurred a long time beforehand.

In Patagonia, the split-off resulted in a peculiar block movement described by A. Windhausen [80] as follows: "The new upheaval began with vast regional movements during the Middle Cretaceous," so that the surface of Patagonia "was changed from a region of pronounced declivity to an overall depression, subject to arid or semi-arid conditions and covered with stony wastelands and sandy plains."

If we continue our comparison of opposite Atlantic coastlines farther north, we find that the Atlas range, which is situated on the northern border of the African continent, and whose folding took place chiefly during the Oligocene, but had already begun in the Cretaceous, has no counterpart on the American side.[15] This agrees with our assumption put forward in the reconstruction that the Atlantic rift had been open for a long period already in this area. In fact, it is possible that here also the rift had been nonexistent at one time, but the start of the split-off must have occurred before the Carboniferous. Further, the great depth of the ocean in the western part of the North Atlantic perhaps means that the sea floor is older here. One should also note the contrast between the Iberian peninsula and the opposite American coastal region, a contrast which makes it most unlikely that the coastlines were formerly in direct connection. In any case, however, drift theory would not imply such a view, because between Spain and America there lies the broad submarine massif of the Azores. As I attempted to prove from the earliest transatlantic echo-sounding profile [37], this massif probably represents a layer of detritus composed of continental material whose original extent can be estimated as possibly 1000 km or more.

The geology of these islands, as well as that of the other Atlantic islands, supports the idea that they are really fragments of continents.

[15] Gentil, and recently Staub also [214], would like to see such a continuation in the ranges of Central America—especially the Antilles—which are the same age; but Jaworski has objected that this is incompatible with the generally accepted theory of Suess, whereby the eastern cordilleran arc of South America is continued as the Lesser Antilles and thus curves westwards again without sending out eastward spurs.

(The question remains, of course, whether a large part of their substructure and that of the mid-Atlantic submarine ridge in general may not be basaltic.)

Gagel [81] also came to the conclusion that the Canaries and Madeira "are split-off remains of the European–African continent, and first separated from it in relatively recent times."

In the area of the Greater Antilles, Matley recently made a geological examination of the Cayman Islands [105] and arrived at the conclusion that the situation there could best be explained by drift theory: "First, all the islands of the Greater Antilles group, though frequently separated by considerable distances and depths of ocean, have a remarkably close family likeness in character, facies and correlation between geological formations and volcanic rock series. Their geological history, so far as it is known, is also very similar. These factors are not only not unfavourable to the view that these islands were once much closer than they are today, but on the contrary, are corroboration of the theory. Moreover, the great submarine depressions of the Caribbean Sea, such as the Bartlett trench (between Jamaica and Cuba), which Taber already claimed to be a trough fault, are so deep that it is difficult to understand how subsidences in the Antillean land mass could have sunk so far into the crust." This is only a minor detail, it is true, but from such small pieces of mosaic the large-scale picture of the whole surface of the earth is finally assembled.

Farther north still, we find in direct sequence three ancient fold zones which carry over from one side of the Atlantic to the other, and once again provide very remarkable confirmation of the idea of direct connection at one time.

What most strikes the eye is the Carboniferous folds which E. Suess called the Armorican mountains and which make the North American coal fields seem to be the direct continuation of the European. These mountains, now much levelled-off, come from the continental inland region of Europe and extend first towards the west-northwest in an arcuate formation, then west to form a wild, irregular (so-called "ria") coastline in southwest Ireland and Brittany. The southernmost folded ranges of this system, which cross France, appear to turn completely south in the offshore continental shelf and to continue on the Iberian peninsula on the other side of the deep-sea rift of the Bay of Biscay, which is shaped like an opened book. Suess called this offshoot the "Asturian swirl." The main mountain chains, however, obviously continue through the northern part of the shelf in a westerly direction, though the tops have been eroded down by the breakers, and they point out to the Atlantic here, as though insisting on a continuation westwards.[16]

This continuation on the American side is formed, as Bertrand was the first to discover in 1887, by the offshoots of the Appalachians in Nova Scotia and southeastern Newfoundland. Here also a Carboniferous range of fold mountains ends, folded in a northerly direction as in Europe; this produces a ria coastline, and the range probably crosses the shelf of the Newfoundland Bank. Its direction, elsewhere northeasterly, turns directly eastwards near the rift area. According to the ideas held to date, it was already assumed that a single large fold system was involved, described by E. Suess as the "transatlantic Altaides." Drift theory simplifies matters considerably: in the reconstruction based on the theory, the two components are brought into virtual contact, whereas up till now a sunken middle section had to be assumed, longer than the terminal sections as known to us—a difficulty that Penck had already experienced. On the junction line of the rift lie some sporadic elevations of the sea floor, regarded hitherto as peaks of the sunken chain. Our theory sees them as fragments of the edges of the separating blocks, whose detachment is easily understandable in just such a region of tectonic disturbance.

In Europe, continuing immediately to the north, extend the fold mountains of an even older range formed between the Silurian and the Devonian, which passes through Norway and northern Britain. E. Suess named this the Caledonian Range. Andrée [83] and Tilmann [84] have taken up the question of the continuation of this range of folds in the "Canadian Caledonians" (Termier), i.e., the Canadian Appalachians, which were already folded by Caledonian times. Naturally, the correspondence between the two systems is not affected by the fact that this Caledonian fold system in America was once again altered by the previously discussed Armorican folds; in Europe the process took place only in the central area (the Hohes Venn and the Ardennes), but not in the north. The abutting sections of these Caledonian folds should be sought in the Scottish Highlands and Northern Ireland on the one side and in Newfoundland on the other.

[16] Kossmat's view [82], which differs from that of E. Suess, is that all the European folds bend around in the ocean regions and return towards the Iberian peninsula; this view would be difficult to support because so large a curve of folds could not be contained within the shelf.

Again just north of the Caledonian fold system in Europe lies the still older (Algonkian) gneiss ranges of the Hebrides and northern Scotland. The corresponding system on the American side of the Atlantic is the gneiss mountains of Labrador, which are of the same age; they extend right up to the Strait of Belle Isle in the south and far into Canada. In Europe, the strike is northeast–southwest; in America it changes from that direction to east–west. Dacqué [22] noted here: "From this one can infer that the chain reached across the North Atlantic Ocean." The alleged sunken link must have been 3000 km long according to previous conceptions, and the straight-line projection of the European section to the American is several thousand km in the direction of South America, taking the present-day position of the continents. According to drift theory, the American land mass once again must be displaced sideways and rotated so that it joins directly onto the European continent and appears as an extension thereof (in the reconstruction picture).

In the region we have just considered, there are also the terminal moraines of the great Pleistocene inland icecaps of North America and Europe. They were deposited at a time when Newfoundland had already been split off from Europe, while in the north near Greenland the blocks were still joined. In any case, North America must at that time have been much closer to Europe than today. If one considers the moraines in our reconstruction, which holds for the period before separation, they join up without gaps or breaks, as shown in Figure 19; this would be highly improbable if the coastlines at the times of their deposition had already been separated by their present-day distance of 2500 km especially since the American end today lies 4½ degrees of latitude farther south than the European.

The coastal correspondences either side of the Atlantic have already been discussed, namely the folds of the Cape mountains and the sierras of Buenos Aires, the conformity of volcanic rocks, sediments, strike directions and countless other details of the huge gneiss plateaux of Brazil and Africa, the Armorican, Caledonian and Algonkian folds and the Pleistocene terminal moraines. Even though the theory in certain individual cases may still be uncertain, the totality of these points of correspondence constitutes an almost incontrovertible proof of the correctness of our belief that the Atlantic is to be regarded as an expanded rift. Of crucial importance here is the fact that, although the blocks must be rejoined on the basis of other features—their outlines, especially—the conjunction brings the continuation of each

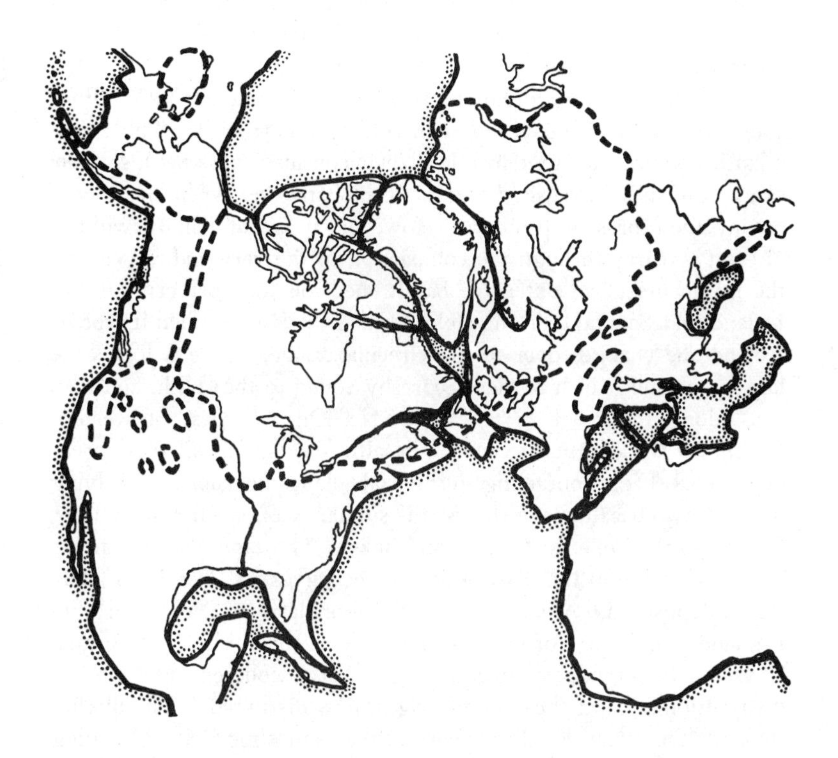

FIG. 19. Boundaries of the Quaternary inland ice, entered on the reconstruction for the period before North America was detached.

formation on the farther side into perfect contact with the end of the formation on the near side. It is just as if we were to refit the torn pieces of a newspaper by matching their edges and then check whether the lines of print run smoothly across. If they do, there is nothing left but to conclude that the pieces were in fact joined in this way. If only one line was available for the test, we would still have found a high probability for the accuracy of fit, but if we have *n* lines, this probability is raised to the *n*th power. It is certainly of some value to make the significance of this clear. Let us assume that we can bet ten to one on the correctness of drift theory just on the first "line" alone—the folding of the Cape mountains and the sierras of Buenos Aires; then, since there are at least six such independent tests available, we can bet a million ($10^6$) to one on the theory being right, in view of our knowledge of the six tests. These figures may be regarded as exaggerated, but they should show the significance of a plurality of independent tests.

North of the region discussed so far, the Atlantic rift forks either side of Greenland and becomes increasingly narrower. The transatlantic conformities therefore lose their conclusiveness because their origin becomes increasingly easy to explain even when the blocks occupy their present position. However, it is not wholly without interest to carry through the comparison to the very end. We find the fragments of an extensive basalt sheet on the northern edge of Ireland and Scotland, on the Hebrides and the Faroes; then it crosses Iceland and changes over to the Greenland side, where it forms the large peninsula which borders Scoresby Sound to the south, and then runs along the coast as far as 75° N. On the coast of western Greenland we also find extensive basaltic sheets. In all these areas occur coal beds containing terrestrial plant inclusions and lying between two basaltic lava sheets; the similarity of one area to another has led to the belief in former land links. The same conclusion has been derived from the distribution of the terrestrial Devonian "Old Red" deposits in America from Newfoundland to New York, in England, southern Norway, the Baltic, Greenland and Spitsbergen. These finds provide an overall picture of a coherent and unified distribution zone at the time of origin, now disrupted. The disruption was the result of submergence of the connecting links, according to previous conceptions, or the result of break-up and relative displacement, according to drift theory.

It is worth mentioning here that similar unfolded Carboniferous beds occur in northeastern Greenland on the one side (81° N) and on the other at Spitsbergen.

Furthermore, the anticipated correspondence in structure holds between Greenland and North America. According to the U. S. Geological Survey's "Geologic Map of North America," there are many pre-Cambrian intrusive rocks in the gneiss near Cape Farewell and northwest thereof, found again in exactly the corresponding places on the American side, i.e., on the northern side of the Straits of Belle Isle. At Smith Sound and Robeson Channel in northwestern Greenland, the displacement does not involve pulling the rift edges apart, but a horizontal large-scale dislocation, a so-called strike-slip, or lateral fault. Grinnell Land slid along Greenland, whereby the remarkably linear borders of the two blocks were produced. This displacement can be seen in the section of Lauge-Koch's [85] geological map of northwestern Greenland shown in Figure 20, if one looks for the boundary line between the Devonian and the Silurian periods; this lies at 80° 10′ in Grinnell Land and 81° 30′ in Greenland. Also, in the Caledonian fold system, discovered by this author, which stretches across from Greenland to Grinnell Land, one can detect the same displacement.

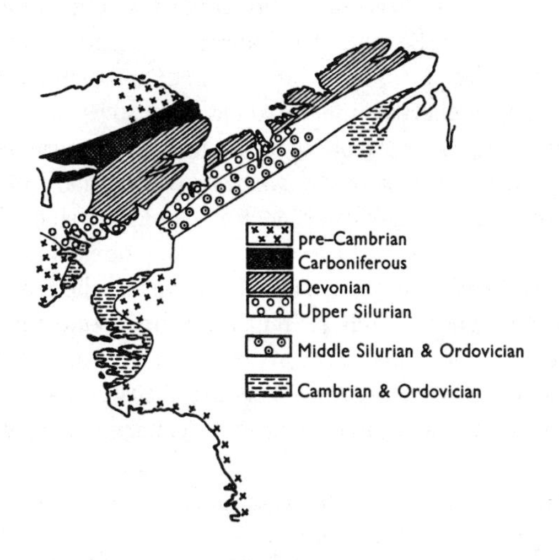


FIG. 20. Geological map of northwestern Greenland, according to Lauge-Koch.

At this point, some further brief indications should be made of how we went about constructing the pre-Atlantic continental connections. We shall give later on a more comprehensive account of the relevant phenomena, such as plasticity of the sial blocks, underground fusion processes and so on; but to avoid misconceptions it is necessary to say something about these things even when only comparing the rift margins on a geological basis.

In North America, our reconstruction shows some deviation from today's map insofar as Labrador seems to be pushed appreciably northwestwards. It was assumed that the strong pull which finally tore Newfoundland from Ireland produced an extension and superficial tearing of the region common to both blocks before actual rupture. On the American side, not only was the Newfoundland block (including the Newfoundland Bank) broken off and rotated by about 30°, but the whole of Labrador was given the opportunity to drop to the southeast, so that the previously straight trough fault formed by the St. Lawrence River and the Strait of Belle Isle took on its present-day S-shaped configuration. Further, the shallow waters of Hudson Bay and the North Sea may have had their origin, or been enlarged, during this pulling process. Therefore the Newfoundland shelf suffers a double positional correction, a rotation and a thrust to the northwest; this reconstruction matches it more closely with the shelf line of Nova Scotia, whereas it now extends far beyond it.

Iceland is assumed to have lain between a double rift, and the present-day depth chart of the surrounding waters seems to indicate this is true. Perhaps a trough fault first arose here between the gneiss massifs of Greenland and Norway which then partly filled up with molten sial from below the block. However, since the rest was filled with sima, as the Red Sea is today, a renewed compression of the block could have had the effect of cutting off the sima filler from the deeper zone and extruding it upwards to create the large-scale inundations of flood basalt. It seems very reasonable to suppose that this in fact took place in the Tertiary, because as a result of the westward drift of South America in the Tertiary a torque must have been transmitted to North America that would have appeared as a compression towards the north so long as the Ireland–Newfoundland chains were acting as an anchor.

We should also give brief consideration to the mid-Atlantic ridge in this connection.[17] Haug considered this as the beginning of the folding of a gigantic geosyncline comprising the whole Atlantic Ocean region, but today this conception is generally recognized as inadequate. We refer to Andrée's critique [16] only. In my opinion we are dealing here, in any case, with a by-product of the separation of the blocks. One can assume that, instead of a single rift, a network of fissures was formed and hence a layer of detritus, most of which sank below sea level as the substratum moved away and flattened out. The detrital layer zone may have been very widespread in places where the present-day margins no longer match closely.

We said previously that the Azores area corresponds to a detrital layer which may have been more than 1000 km wide originally, so far as we can estimate. This is of course an exceptional case and in most regions the mid-Atlantic ridge is much narrower. From the

[17] Cf. the chart of the Atlantic Ocean in Schott's *Geographie des Atlantischen Ozeans*, 2nd ed., Hamburg, 1926.

map drawn by du Toit (Fig. 18), one would deduce on the basis of the present-day marginal shelf a detrital layer of only a few hundred km width, and it may have been still narrower in places; this agrees with the fact that the block edges are, even today, remarkably congruent, if one excepts a few misfits such as the Abrolhos Bank or the salient area at the mouth of the Niger. Our reconstruction maps in Figures 4 and 5 are diagrammatic in so far as they pay insufficient attention perhaps to this detrital belt, which is hard to assess. However, whether it will ever be possible to carry out the reconstruction exactly in such detail must remain undecided at the present time, because even if the profile of the Atlantic floor were fully and accurately known, it would still be uncertain how much of it is basalt and once lay underneath the two present-day continental blocks and only during the separation process was torn out by extraction of the subcontinental material, or flowed out. We were unable to take this part into account when making the reconstruction.

Geologically speaking, there is less to be said on the subject of the other continental connecting links that we have assumed than on the Atlantic rift.

Madagascar, like its neighbour Africa, consists of a plateau of folded gneiss with a northeasterly strike. The same marine sediments have been deposited both sides of the rift line, implying that since the Triassic both bodies of land have been separated by an inundated trough fault; the Madagascan terrestrial fauna also makes this a necessary assumption. However, Lemoine [87] states that in the Middle Tertiary, when India had already been detached, two animals, the potamochœrus and the hippopotamus, immigrated from Africa. In the opinion of Lemoine, these could only have swum at most an inlet 30 km wide, whereas now the Mozambique Channel is a good 400 km across. Therefore it is only after this time that the Madagascan block could have undergone a submarine breakaway from Africa, which explains the considerable lead that India had over Madagascar in the drift to the northeast.

A factor of considerable importance in the structure of Africa is the rifts, which run mostly north–south and are found particularly in eastern Africa. In an interesting investigation of "regions of tension" of the earth [107], Evans has stressed many points in favour of drift theory, especially the following: "Much of the structure of the African continent has yet to be determined; but, so far as it is known, it appears everywhere to support the view that there is evidence of the prevalence of tension directed outwards from the centre. This is in accordance with Wegener's contention that at the beginning of Mesozoic times there was a great 'Ur-Kontinent,' of which Africa was the centre, and that it has since been broken up by a relative movement of South America to the west, of West Antarctica to the southwest, of India to the northeast, of Australia to the east, and East Antarctica to the southeast."[18]

India also is a flat plateau of folded gneiss. The folding still shows its morphogenic effect in the ancient Arvalli Mountains in the extreme northwest at the edge of the Great Indian Desert (Thar), and again in the Korana Mountains, also very ancient. According to Suess, it runs N 36° E in the former, northeast in the latter. Both directions thus correspond closely enough to the African and Madagascan strike directions, and all the more closely when India is rotated by the small amount required for our reconstruction. Besides this there is also a somewhat younger but still ancient fold in the Ghats of Nellore or in the Vellakonda Mountains, with a north–south strike very comparable with the north–south strike in Africa, which is also more recent. The presence of diamonds in India is connected with their presence in South Africa. It is assumed in our reconstruction that the west coast of India was joined to the east coast of Madagascar. Each coastline consists of the remarkably straight edge of a rift in a gneiss plateau, suggesting that after the split they could have slid alongside each other like Grinnell Land and Greenland. At the northern end of this rift, about 10° of latitude long on both coastlines, basalt occurs on each side. In India, this is the basalt sheet of the Deccan which begins at 16° N; it originated at the onset of the Tertiary, and therefore may perhaps be causally connected with the separation of the two regions. In Madagascar, the northernmost part of the island is wholly made up of two basalts of different age, though apparently of still undetermined date.

The vast folds of the Himalaya, essentially of Tertiary origin, signify the compression of a large portion of the earth's crust; the periphery of the continent of Asia looks quite different if these folds are subject to our type of reconstruction. Probably all eastern Asia via Tibet and Mongolia up to Lake Baikal, and even to the Bering Strait perhaps, took part in this compression. The latest investigations have shown that the recent folding processes were by no means

[18] When these movements began, the cardinal points of the compass were significantly different because of the different position of the poles.

restricted to the Himalaya itself; for example, in the Peter the Great Mountains, Eocene layers were uplifted by folding to 5600 m above sea level, and in the Tien Shan Mountains large overthrusts were produced [88]. However, even where such fold phenomena are absent, the recent uplift of unfolded land is also closely connected with this folding process. The great masses of sial, which subside into the depths when folding occurs, must melt there and, as they diffuse outwards, support the adjacent blocks and lift them. If we limit ourselves to considering the highest region of the Asiatic block, which lies on an average about 4000 m above sea level and which measures 1000 km in the direction of the compression; and if we assume (despite the greater altitude) only the same foreshortening as in the case of the Alps, i.e., to a quarter of the original extent: then we obtain a 3000-km displacement of India, so that before the compression it must

FIG. 21. The Lemurian compression.

have been alongside Madagascar. There is no room here for a sunken Lemuria in the old sense of the term.

The traces of this gigantic compression can also be observed both right and left of the rather narrow compression zone. The detachment of Madagascar from Africa and the whole system of recent trough faults in eastern Africa, to which both the Red Sea and the Jordan valley belong, are aspects of the same picture. The Somali peninsula may have been pulled somewhat northwards and the compression which formed Abyssinian ranges may be related to this; the sial mass which was here forced downwards across the fusion isotherm flowed under the block to the northeast, to well up in the angle between Abyssinia and the Somali peninsula. Arabia also experienced the northeasterly pull and the offshoots of the Akdar Mountains penetrated the Persian chains like a spur. The fan-like formation of the Hindu Kush and Sulaiman ranges indicates that here the westerly limit of the compressive thrust is reached; its faithful mirror image occurs on the eastern margin of the compression, where the mountain ranges of Burma were pulled away from the direction traced out through Annam, Malacca and Sumatra and were pointed north–south. The whole of eastern Asia was probably affected by this compression, whose western boundary lies in the stepped folds between the Hindu Kush and Lake Baikal and their continuation up to the Bering Strait; the eastern limit is formed by the convex littoral and the island arcs of eastern Asia.

At first sight, these ideas might seem fantastic, but they are completely confirmed by recent investigations of research workers in the field of mountain structure. This applies particularly to the large-scale investigation by Argand [20] on the structure of Asia, which appeared in 1924.

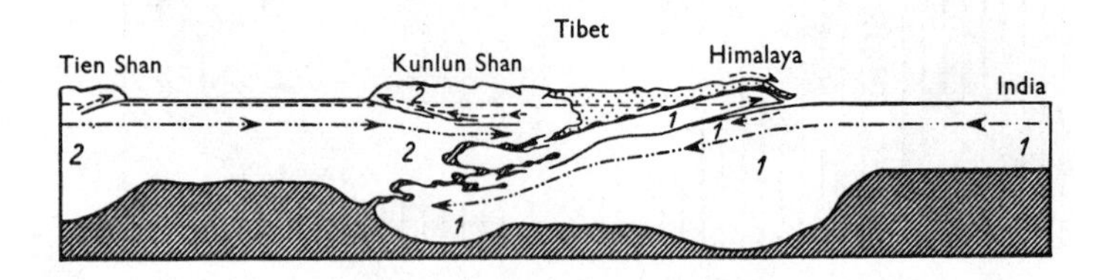


FIG. 22. Longitudinal section through the Lemurian compression, according to Argand.

1 = Lemuria (India); 2 = Asia.

In Figure 22 we reproduce one of Argand's diagrams which illustrates his concept of the vast compression of upland Asia; it represents a longitudinal section from India up to the Tien Shan range as he thought it would have been at the close of the Tertiary. The hatched areas signify the supporting sima; the unhatched, the sial blocks; the dotted, the products left by the Tethys. The basic (sima) rocks entrained by the sial are indicated. The arrows show the relative movement. Overall we are involved here with a gigantic overthrust, in which the sial Lemurian block was forced under the Asiatic block.

Of the other illustrations in this important work we reproduce only that shown in Figure 23; it will clearly demonstrate how completely the results obtained by this prominent structural geologist agree with those of drift theory. Argand draws attention to the following

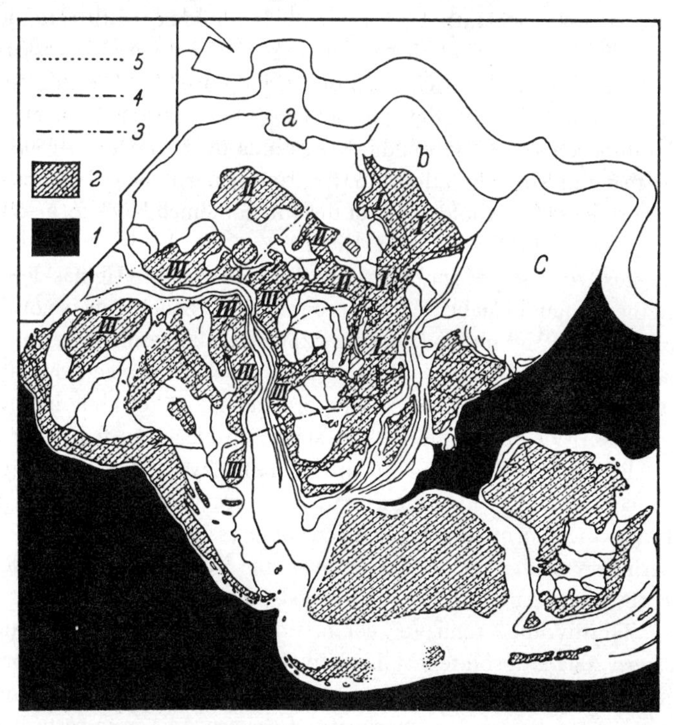


FIG. 23. Tectonic chart of Gondwanaland, according to Argand.

1 = predominantly sima; 2 = areas where anticlinal sial folding preponderates; I, II, III = the three branches of the inner virgation of the Gondwanaland block; 3 = ridges of sial fold; 4 = trough lines of sial fold; 5 = joint lines; *a, b, c* = African, Arabian and Indian foothills of Gondwanaland.

peculiarity, amongst others. If one considers the three outlines of the sial fold regions I, II and III, regarded by Argand as a large-scale virgation, the individual shapes are curved in a way similar to the South American Andes but the curvature decreases as we go eastwards. Argand concludes (pp. 317, 318): "A plastic deformation occurred due to a thrust that came from the west and was transmitted to the whole framework of Gondwanaland; this thrust made itself felt right through the continental land mass and its effect on the configuration slowly faded towards the east." As in the case of all such sial folds, friction in the underlying sima and the internal deformation of sial must be taken into account in the explanation; Argand says that in this case there was, "before the Atlantic rift occurred, the resistance of the Pacific sima to the westward drive of Gondwana, i.e., on the leading edge of what is now South America. . . . It would be fruitless to look for any explanation of all these homologies without such stress relationships between the Andes and this virgation. . . . The existence of Andean movements north from the Tanganyika District, evidenced by the discordance of the middle Cretaceous over the Jurassic beds, shows that this stress relationship, far from being illusory, embraced at the very least the whole width of the still-combined block formed by what is now South America and Africa."

We must refer to yet another result due to Argand. He has determined the amount of sialic folding of the chief fold zones; this is not the place to discuss the method, but he expressed the result in terms of tonnage per unit distance. He also distinguished between the sial-fold tonnage and that of the "newly formed chains," of little importance for energy considerations. By statistical means he finds that in the mediterranean fold range (Alps to Himalaya), the "tonnage" varies considerably, in contrast to the circum-Pacific folds. In particular, the enormous central Asian compression has no parallel on the periphery of the Pacific. Moreover, the "tonnage" on the west coast of North America greatly exceeds that on the east coast of Asia. Thirdly, the "tonnage" comprising the recent mountain chains of eastern Asia is absolutely and relatively greater than that in North America, where new-range tonnage is almost completely lacking; this further emphasises the inferiority of eastern Asia with respect to the amount of folding.

The first result—high degree of variability in extent of folding in the mediterranean fold zone—is explained by Argand as a consequence of the inhomogeneity of the sial blocks found here. He says: "Conversely, the slight variations of tonnage in the circum-Pacific region indicate the presence or prevalence of a more homogeneous and yielding material under the Pacific Ocean than that which composes the very heterogeneous and always inflexible continental blocks. . . . Drift theory has no difficulty in taking into account the facts of tonnage distribution and their direct interpretation. The theory holds that the relatively homogeneous and compliant material under the Pacific is sima. . . . Drift theory easily explains the second and third groups of facts, which are expressive of the energy deficiency of eastern Asia compared with America. This theory allows for processes at the block front, in which the sial, under certain conditions, thrusts against the sima and folds; and at the rear, which involve a withdrawal of sial, causing a more or less complete interruption in fold formation, plus the effects of tensile stresses: lateral fractures; button-hole-like tear faults, which form marginal inlets; the leaving behind of mountain chains which thenceforth are entrained in the track of the continent as more or less detached island arcs; while the sima, now forced to adapt to new conditions, is elevated behind the block. Due to the time delay involved in the complete uplift of the sima, deep troughs arise from which derives the classical concept of foredeeps. Since the drift theory requires that the first type of process should have taken place predominantly on the western margin of America, and the second type over a long period of time in eastern Asia, the superiority of tonnage of the former over the latter is self-explanatory."

Argand adds: "The elegance with which drift theory explains these significant facts, which were not known when the theory was originated, is certainly a strong point in its favour. Strictly speaking, none of these facts really proves drift theory or even the presence of sima, but they all fit in excellently with both ideas to an extent that makes them highly probable."

We have now given the views of Argand, who may be seen to have considered the main outlines of the whole globe in his work on the structure of Asia.

It would also be well worth while to make an exact geological comparison between the east coast of India and the west coast of Australia, because these littorals or, better, shelf margins, were joined together up to about the Jurassic, according to our theory. However, no such comparison has apparently been made so far, at least from the geological viewpoint. The eastern littoral of India represents a precipitous break in the gneiss plateau. This pattern is interrupted only by the narrow, trough-shaped coal field of the Godavari, composed of the lower Gondwana beds. The upper Gondwana beds follow the coastline and lie unconformably right across its end. Western Australia also forms a gneiss plateau with an undulant surface like that of India and Africa. This plateau drops down to the sea along the coastline with a long, steep edge, the Darling range and its northerly continuation. In front of the steep edge lies a flat subsidence, an area composed of Palæozoic and Mesozoic strata with a few basalt intrusions; in front of this again at the coast is a narrow stretch of gneiss, which at times disappears altogether. The sediments mentioned also contain coal where the Irwin River runs. The strike of the gneiss folds in Australia is everywhere north–south; if joined on to India, the strike would be changed to northeast–southwest, that is, parallel to the main strike direction in India.

In eastern Australia, the Australian cordilleras (essentially a Carboniferous fold system) follow the coast from south to north, ending in a fold system which recedes westwards stepwise; the individual folds always run precisely north–south. Just like the stepped folds between the Hindu Kush and Lake Baikal, the Australian cordilleras indicate the lateral boundary of the compression; the vast folds of the Andes, which start in Alaska and stretch throughout four continents, end here. The most westerly chains of the Australian cordilleras are the oldest, the most easterly being the most recent. Tasmania forms a continuation of this fold system. It is interesting here to observe in the structure of the range the mirror-image similarity to the South American Andes, where the most easterly chains are the oldest, because they lie on the other side of the Pole. The most recent chains are missing in Australia, but Suess [12] considers that they occur again in New Zealand. Of course, the folding process did not extend even here into the Tertiary period: "According to most New Zealand geologists, the main folding which formed the Maorian range took place between the Jurassic and the Cretaceous." Before this, virtually everything was covered by sea, and it was the folding process which first "turned the New Zealand area into a land mass." Upper Cretaceous and Tertiary beds are mostly unfolded border areas. In fact, Cretaceous deposits on South Island are found only on the east coast, not on the west. During Tertiary times occurred the "West-coast breakaway," "because Tertiary marine deposits are also found there." In the Upper Tertiary, finally, there arose other folds, faults and overthrusts, on a reduced scale of course, which gave the mountains the form they have today (Wilckens [89]). Drift theory explains this whole matter by saying that New Zealand at one time was part of the eastern edge of the Australian block, so that its main folding process was connected with that of the Australian cordilleras. When the New Zealand chains were detached as an island arc, however, the folding process ceased there. The Upper Tertiary disturbance may well be connected with the fact that the Australian block moved past and drifted away from New Zealand.

Many details of these movements of Australia just referred to are revealed to us in particular by the depth chart around the New Guinea region. As shown schematically in Figure 24, from the southeast

FIG. 24. Dispersal of island arcs by New Guinea; diagrammatic.

came the huge Australian block with its front thickened like an anvil: New Guinea (folded to form a young high-altitude mountain range) plus shelf; this forced itself between the chains of the southernmost Sunda Islands and the Bismarck Archipelago. In the depth chart shown in Figure 25,[19] consider the two most southerly groups of the Sundas; the Java–Wetar chain running west–east curves up at the end to form a spiral round the Banda islands to the Siboga Bank following a northeast, north, northwest, west and finally a southwest course. The Timor island chain which lies south of the Java–Wetar has a distorted and variable course which is evidence of collision with the Australian shelf; H. A. Brouwer [90] has given the detailed geological reasoning behind this. This chain also is violently twisted in a spiral, like the Java-Wetar one, up to Buru. An interesting detail should be brought up here to which Brouwer devoted a special paper [112]: the inner island chain is studded with volcanoes,

[19] This is shown most clearly in the first-rate map of the Sunda Islands in G. A. Molengraaff's "Modern Deep-Sea Research in the East Indian Archipelago," *The Geographical Journal*, Feb. 1921, pp. 92–121, which gives elevations and soundings at equal intervals.

FIG. 25. Depth chart for the waters surrounding New Guinea.

generally still active today; only on the stretch between the two islands of Pantar and Damar (exclusive) are the once-active volcanoes extinct. That is just the part, however, against which the outer chain at the north border of Timor was pressed by the Australian shelf, so that here the bending process was halted, though it continued everywhere else. These facts fit excellently with the idea of a collision with the Australian block and are also very instructive in connection with the problem of the origin of the volcanoes through the pressure which arose from the bending of the island chains.

One can see a very interesting supplement to this collision process on the east side of New Guinea: the latter moved up from the southeast and scraped against the islands of the Bismarck Archipelago, whereby the island of New Britain (Neupommern) was gripped by the portion that once formed its southeast end and entrained, so that the long island was rotated more than 90° and bent into a semicircle. A deep trough remained behind the trailing edge, showing the power involved in the process, since the sima has still not been able to fill it.

It will seem to many people rather rash to deduce such conclusions just from a depth chart, but this source almost always turns out to be a reliable guide to the movement of blocks, particularly for the more recent periods.

There are also many individual phenomena in the Sunda Archipelago which point to the validity of our ideas. For example, Wanner [96] explained the deep ocean between Buru and Sula Besi, which would not be expected from the structure of that region, on the grounds that Buru has been displaced 10 km in a horizontal direction; this dovetails well with our theories. G. A. F. Molengraaff [97] gives a chart of the Sunda Islands showing the region where coral reefs have been elevated by more than 5 m. This region coincides astonishingly closely with the one where, according to drift theory, the sial must have been thickened by compression; this is the whole region lying just north of the Australian block (apart from the southwest coasts of Sumatra and Java), right up to Celebes (Sulawesi) inclusive, as well as the north and northwest coasts of New Guinea. According to Gagel [98], there are quite recent terraces, which have been uplifted 1000, 1250, perhaps almost 1700 m, in New Guinea at Cape König Wilhelm, and according to Sapper [99] they are also to be found in New Britain. This very remarkable phenomenon means, whatever else may be involved, that here and in very recent times, powerful forces were applied, and that fits our idea that a regional collision occurred.

Since just here in the Sunda Islands the consequences of drift theory seem so fantastic at first sight, it is certainly worth noting that the Dutch geologists who are working in the Sunda archipelago were among the first to take their stand on drift theory; the very first of these was Molengraaff, who came out for the theory as early as 1916 [91]; later also van Vuuren [92], Wing Easton [93], Escher [95] and recently Smit Sibinga [94] in particular, who has given a complete presentation of the geological development of the Sunda Archipelago from the standpoint of drift theory, and in doing so has also solved the old problem of the origin of the peculiar shapes of Celebes and Halmahera. He concludes: "The small Sunda Islands, Celebes and the Moluccas represent marginal chains originally cut off from the Sunda land mass; at first they formed an ordinary double chain, but afterwards took on their present shape due to a collision with the Australian continent." We give here the concluding section of his investigation:

"In a final section we would like to indicate point by point a few geological facts and peculiarities relating to the Molucca Archipelago that are explained, or explained better, by the working hypothesis based on the fundamental ideas of Taylor and Wegener, as developed above, than by any other theory.

"1. There is no need in drift theory for a submergence of former land masses under the ocean to explain the present-day relief, the process of mountain building and the disappearance of former land bridges; that is, drift theory is in agreement with isostasy theory.

"2. Drift theory explains the present-day configuration in an unambiguous and logical manner by a collision between a Molucca chain (originally double) and the Australian continent.

"3. It provides an explanation for the odd S-shape of the northern arm of Celebes, very unusual and puzzling for a geanticline. This also came about as a result of pressure from the Australian continent, which displaced the Timor–Ceram chain as far as Celebes and thereby broke (crushed) the chain in between Buru and the Sula group.

"4. Drift theory provides an unforced solution for the remarkable form of the island chains that circle the Banda Sea basin, explaining them as a 'compressed chain.' We have already discussed in detail above the untenable consequences to which the contraction theory leads in this problem.

"5. Drift theory explains the divergence of the transverse faults in the Timor–Ceram chain from the Banda basin outwards as a consequence of the fact that this chain was caught in the thrust of the Australian continent, a phenomenon inexplicable from the standpoint of contraction theory.

"6. Drift theory enables one to understand the anomalous Tertiary strike directions of the outer chain, since they would have arisen while the chain still had its original shape, before compression.

"7. Drift theory allows the orogenic force to have come from the Australian[20] continent. This explains why it was precisely the outer chain which came into direct contact with this continent—that was so much more strongly folded and overturned than the inner chain, Celebes and the Halmahera group. The inner chain never came into contact with Australia; these orogenic forces were only transmitted to Celebes via the outer chain and must therefore have lost in intensity; the Halmahera group had almost the same intimate contact as that between Australia and the outer chain. If, on the contrary, one assumes a tangential pressure coming from the Banda basin, one would then expect the most intensive orogenesis in the inner chain and eastern Celebes.

"8. In explaining the formation of mountains, drift theory avoids the idea of a primeval land with heterogeneous geological and zoological elements.

[20] The original has "Asiatic," doubtless a misprint.

"9. In the rupture of the outer chain between the Tukang Besi and Banggai Islands and the consequent relief of the stresses, drift theory can find an explanation for the interruption in mountain building during the Lower Pliocene, the process then restarting, even though with less intensity, when the contact with Celebes was made during the Upper Pliocene.

"10. Drift theory offers an acceptable explanation for the striking geological difference between Celebes west, and Celebes east of the Boni-Posso depression. The extinction of active vulcanism in central Celebes and its renewed onset in the northern arm of the island can be explained in the same manner as the break in active vulcanism between Pantar and Damar (Brouwer), that is to say, by the penetration of the outer chain (eastern Celebes) into the inner (western Celebes).

"11. The stratigraphic pattern of the eastern part of the East Indian Archipelago becomes clearer and more distinct. Since the most recent Palæozoic times, up to Neocene times, an intermittent transgression penetrated further and further into the Sunda land areas, closely associated with a simultaneous formation and detachment of marginal island chains. From a geosynclinal strip, which lay in front of the edge of the Mesozoic Sunda land mass, the outer chain developed; from another strip, which lay in front of the Tertiary Sunda land, the inner chain developed in the Early Miocene period, the marginal chains—folds in a geosyncline that was chiefly a Neocene formation—still remaining united with the Sunda land mass.

"12. Drift theory makes it possible to give a more satisfactory explanation of the distribution of fauna in the Moluccas; it requires there to have been a former land link between the Philippines, the Moluccas and Java, and one between the Halmahera group and northern Celebes, which is just what zoogeographers believe."

One can now see that drift theory has already become a tool of the professional geologist for this very puzzling region of the earth.

Two submarine ridges join New Guinea and northeastern Australia to the two islands of New Zealand, and appear to indicate the path of the displacement; the ridges may be former land areas which were flattened by traction and therefore submerged, but in part also, they may be fused remains of the underside of the block.

On the subject of links between Australia and Antarctica, our ignorance of the latter continent allows us to say but little. A wide strip of Tertiary sediments runs along the whole southern rim of Australia and extends through Bass Strait; it is only found again in New Zealand, the east coast of Australia being free of it. It may be that, during Tertiary times, an inundated trough fault already separated Australia from Antarctica, or even ocean may have separated them, with the exception of the area around the Tasmanian anchor. It is generally assumed that the structure of Tasmania is continued in Victoria Land, Antarctica. On the other hand, Wilckens [89] wrote: "The south-western arc of the New Zealand fold range (the so-called Otago saddle) appears to be cut off abruptly on the east coast of South Island. This termination is not a natural one, but doubtless stems from a fracture. The continuation of the range can be sought in only one direction, i.e., the Graham Land cordillera or "Antarctic Andes."

We should also mention that the east end of the Cape mountains in South Africa represents a fracture in the same way. According to our (admittedly uncertain) reconstruction of the site of Antarctica, we would have to look for the continuation of the range between the Gaussberg and Coats Land, where the coast is still quite unknown, however.

The aforementioned connection between western Antarctica and Tierra del Fuego serves, geologically speaking, as a model for demonstrating drift theory (Fig. 26). As late as the Pliocene, at least a limited exchange of forms must have taken place between Tierra del Fuego and Graham Land, according to palæontological data relating the two areas; this would only have been possible if both headlands had still been located near the island arc of the south Sandwich Group. Since then they have drifted from there westwards, but their narrow link remains lodged in the sima. In the depth chart[21] one can clearly see how the echeloned chains were torn off *seriatim* from the drifting blocks and then left behind. The south Sandwich Group, right in the middle of the rift area, was the one most strongly bowed by this process. The sima inclusions were extruded by the movement. The islands are basaltic and one (Zawadowsky Island) is still volcanically active today. Furthermore, acccording to F. Kühn [100], the late Tertiary Andean folds are missing over the whole chain of the "South Shetlands arc," while the older folds on South Georgia, the South Orkneys, etc., are well known. These very peculiarities are explained by drift theory, for if in fact the folds in South America and

21 H. Heyde has drawn up a good depth chart of Drake Passage; reproduced by F. Kühn [100]. However, the differences between it and our figure are unimportant.

FIG. 26. Depth chart of Drake Passage according to Groll.

Graham Land were produced by westward drift of the blocks, the folding process must have ended, so far as the South Shetlands were concerned, just at the time when they became stuck in the sima.

In this connection, one could also adduce the Permo-Carboniferous glaciation phenomena, which are to be found everywhere on the southern continents, to substantiate the drift theory, since they—like the Old Red in the northern hemisphere—form the fragments of a unified region, and this is much easier to explain by drift theory than by the concept of sunken continents, because the separation between them is so great today. However, this phenomenon will not be fully discussed until the next chapter but one, since the matter is one of interest mainly to climatologists.

If one surveys the results of this chapter, it is impossible to escape the impression that drift theory can today be regarded as well founded geologically, even in its detailed pronouncements. It is true that there are still many opponents of the theory among today's geologists, and objections have been raised from different sides, by Soergel [35], Diener [108], Jaworski [109], W. Penck [111], A. Penck [110] Ampferer [68], Washington [113], Nölke [114] and many others. However, it can be said in a general way that these objections, so far as they are not just misunderstandings (particularly in Diener's case), mostly involve mere side-issues whose solution would have little significance for the basic concepts of drift theory. It may be permissible perhaps to cite the testimony of Argand [20], who assures us that:

"Since 1915, and particularly since 1918, I have spent a long time testing the degree of authenticity of drift theory, drawing upon the whole atlas of tectonic forms at my disposal, and all the points in opposition to movements that I can see. Therefore, if I lack the time today to substantiate some of my deductions, this does not mean anyone can regard them as rash or unfounded, without stretching a point."

On the question of objections to the theory, Argand has this to say: "The soundness of any theory is no more than its ability to portray the entirety of facts known to date. In this respect the theory of drift of the large continental land masses is in an excellent state of health. When it began it was aiming at the unknown (*elle a visé à l'absolu*); as it developed, it gained much in strength and resources without sacrificing anything at all in the way of logical power; on the contrary, it increased in scope and became more and more in harmony with the conceptions that are generally held. This work of purification and refining is quite perceptible in the publications of Wegener. Drift theory is firmly based on the areas where geophysics, geology, biogeography and palæoclimatology overlap, and it has not been refuted. One must have searched for a long time to find objections to the theory, and also have found some in order to be able to appreciate fully a certain unassailability that distinguishes the theory and which comes from great flexibility combined with a vast range of possibilities for vindication. One thinks he has a crucial objection right in his grasp; one more blow and the whole theory must crumble. Yet nothing crumbles; it is just that one has forgotten a point or two. That is the protean resilience of a flexible world-view.

"It is true that there are a number of objections to the theory, but almost all of them are of the kind that I have mentioned. Of those that have been published or can be imagined, only a very few are sound; these concern only a few subsidiary matters and never, so far, the vital issues."

CHAPTER 6

# *Palæontological and Biological Arguments*

PALÆONTOLOGY, zoogeography and phytogeography also have a significant contribution to make in unravelling the problem of how the earth appeared in prehistoric times; the geophysicist easily finds himself on the wrong track if he does not keep constantly in mind the results provided by these branches of science, in order to check his own.

On the other hand, if biologists are concerned at all with the question of drift, they should make use of the facts of geology and geophysics in forming their own judgements. If they do not, unprofitable mistakes are a constant threat to them. To emphasise this is not superfluous, for, as far as I can see, a large proportion of today's biologists believe that it is immaterial whether one assumes sunken continental bridges or drift of continents—a perfectly preposterous attitude. Without any blind acceptance of unfamiliar ideas, it is possible for biologists to realise for themselves that the earth's crust must be made of less dense material than the core, and that, as a result, if the ocean floors were sunken continents and thus had the same thickness of lighter crustal material as the continents, then gravity measurements over the oceans would have to indicate the deficit in attractive force of a rock layer 4 to 5 km thick. Furthermore, from the fact that this is not the case, but that just about the ordinary values of gravitational attraction obtain over ocean areas, biologists must be able to form the conclusion that the assumption of sunken continents should be restricted to continental-shelf regions and coastal waters generally, but excluded when considering the large ocean basins. It is only by keeping in touch with associated sciences that the study of earlier and present-day distribution of organisms over the globe can throw the full weight of its rich factual resources into the task of discovering the truth.

I have made these fundamental points by way of introduction because they seem to me often insufficiently considered in the biological literature on drift theory that has appeared so far; what is more, this is true even where the author has decided in favour of the theory. Von Ubisch [117, 227], Eckhardt [119], Colosi [118], de Beaufort [123] and others have written survey articles on the position that biologists have taken up vis-à-vis drift theory; they agree in general with the theory but almost always without taking sufficient account of the views raised above. It is therefore not surprising that there are even cases like those of Ökland [116] or von Ihering [122]. The former concluded in his examination of drift theory that the North Atlantic was explained just as well by a sunken continent as by drift; the latter came to the same conclusion about the South Atlantic; and both thought that the idea of a sunken continent was perhaps even preferable. In reality, the question has been quite wrongly formulated. Where the ocean basins are involved, it is not a question whether drift theory or the theory of sunken continents is to be preferred, because the latter idea just does not come into the picture. It is simply a matter of choosing between drift theory and the theory of the permanence of the ocean basins.

For the foregoing reasons, we are justified in counting as favourable to drift theory all biological facts which imply that at one time unobstructed land connections lay across today's ocean basins. The number of such facts is legion. It would be a hopeless undertaking for the layman to cite all the relevant facts, and the scope of this book makes it impossible. Anyway, it is quite unnecessary to do so, because there is a copious specialist literature on the subject, a survey of which has been made by Arldt [11] among others; the results are already established in their general outlines and are as good as universally accepted.

The question is particularly clear in the case of the former land connection between South America and Africa. As Stromer has stressed (amongst other facts), the distribution of the Glossopteris flora, that of the reptile family of the Mesosauridæ and much else force one to assume a former large area of dry land joining the southern continents. Jaworski [109] also examined all the objections, of which there is naturally no lack, and he too concluded: "Everything of a geological nature known about western Africa and South America is in complete agreement with the idea derived from zoo- and phytogeography of both present and past eras, namely, that in former times there was a land connection between Africa and South America in place of the South Atlantic ocean of today." On the basis of phytogeography Engler [126] came to the conclusion: "Taking the whole situation into account, the above-mentioned occurrences of plant types common to America and Africa would be best explained if we could show that there had been large islands or a continental bridge between northern Brazil (southeast of the Amazon estuary) and the Bight of Biafra in the west of Africa, and another connection between Natal and Madagascar, whose continuation in a northeasterly direction towards India separated from the Sino-Australian continent, was already claimed long ago. The many points of relationship between the Cape flora and the Australian also make desirable a connection between Africa and Australia via Antarctica." The last connections seem to have been those between northern Brazil and the Guinea coast: "Furthermore, western Africa has in common with tropical South and Central America the manatee (Manatus), which lives in rivers and shallow, warm seas but cannot cross the Atlantic. This means that there must have been a shallow-water connection in the recent past, and one which extended along the northern coasts of the South Atlantic, between western Africa and South America." (Stromer.)

However, it is von Ihering [122] in particular who produced a fund of evidence for this former land connection in his book *Die Geschichte des Atlantischen Ozeans*. We will not go into the details; but the whole book constitutes an argument for this connection, though it involves the untenable interpretation that this was formed by an intermediate continent "Arch-helenis" between today's continental blocks, whose position is regarded as unaltered since then.[22] The connecting link appears to have been broken, as our Figure 1 (p. 7) shows, just before the middle of the Cretaceous.[23]

[22] So far as I can see, von Ihering's book provides no single positive reason for his violent rejection of drift theory. In particular, I have read through his Chapter 20 ("Two World-views: von Ihering's and Taylor-Wegener's") several times with the best intentions, in order to understand his objections. But I found only a perpetual confusion of continent with continental block and of shallow with deep sea. It would therefore seem that his rejection is based, not on observed facts, which as Köppen [127] has also stressed, in reality fit the theory remarkably well, but rather on insufficient knowledge of the nature of drift theory (cf. my reply to von Ihering's criticism in [128]).

[23] The particulars of this event and of the times when the other connections terminated are obviously not quite the same according to all investigators. I still

The case of the former land connection between Europe and North America provides a less simple picture, as shown by Figure 1; clearly it was repeatedly eliminated or at least obstructed by transgressions. The following table, due to Arldt [11], gives the percentage of identical reptiles and mammals on each side of the connection, and is very instructive:

| | Reptiles % | Mammals % |
|---|---|---|
| Carboniferous | 64 | — |
| Permian | 12 | — |
| Triassic | 32 | — |
| Jurassic | 48 | — |
| Lower Cretaceous | 17 | — |
| Upper Cretaceous | 24 | — |
| Eocene | 32 | 35 |
| Oligocene | 29 | 31 |
| Miocene | 27 | 24 |
| Pliocene | ? | 19 |
| Quaternary | ? | 30 |

The run of these numbers agrees well with our vote-score shown in Figure 1, according to which a land connection is assumed by most experts to have existed in the Carboniferous, Triassic, then only for the Lower, not the Upper Jurassic, but again from the Upper Cretaceous onwards through the Lower Tertiary. The concurrence in Carboniferous times is specially striking, possibly because here the fauna is particularly well known. There are many thorough in-

---

believed, at the time when the second edition of this book appeared, that the conclusion derived from the literature then available to me should be that the connection between South America and Africa had lasted as late as the oldest period of the Tertiary; later on, I was able to satisfy myself that it had already disappeared during the Cretaceous according to most investigators. Individual opponents of the drift theory, who have failed to notice that this insignificant correction had already been taken care of in the third edition, are still clinging to this inaccuracy today, and believe in some odd way that they can confute drift theory thereby. In fact, the question of dating has nothing at all to do with the validity or otherwise of drift theory; it is left entirely to the specialised sciences to solve and serves drift theory only by lending precision to its pronouncements. Even if in the future, as is quite possible, there are still small corrections to be made to the dates involved —large corrections need no longer be feared—there would be no reason to advocate a revision of drift theory.

vestigations relating to both the fauna and flora of Carboniferous times in Europe and North America: those of Dawson, Bertrand, Walcott, Ami, Salter and von Klebelsberg, among others. The last-named [129] has referred in particular to the mutual fauna of the marine interlayers of the coal-bearing bed series from the Donetz through Upper Silesia, the Ruhr district, Belgium and England as far as the west of North America, which is remarkable for its short duration. The identities here are in no way limited to those elements which were distributed over the whole globe. We cannot go into more detail here. The lack of identical forms in the case of the reptiles in the Pliocene and Quaternary is obviously due to the cold, which exterminated the older reptile fauna. From the time of their entry into the earth's history onwards, the mammals show the same picture as the reptiles. The correspondences were specially close in the Eocene. The reduction in the number of correlations in the Pliocene should perhaps be attributed to the inland ice apparently already forming in America at that time. We show here the sketch map due to Arldt (Fig. 27), which gives the distribution of those organisms which

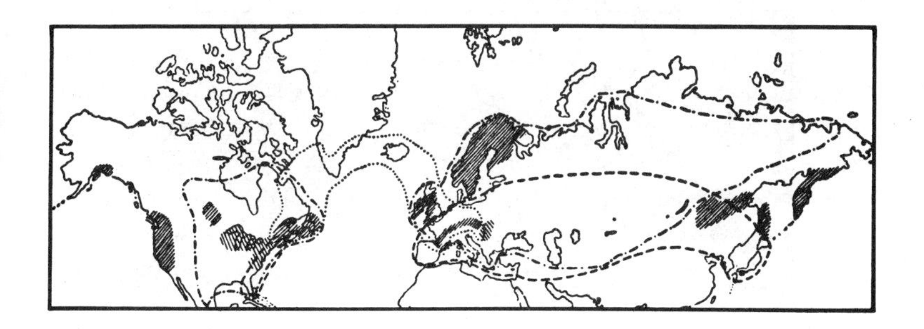

FIG. 27. Distribution of North Atlantic organisms, according to Arldt.

Dotted line = garden snail; dashed = Lumbricidæ (earthworms); dotted and dashed = perch; hatched NE–SW = pearl mussel; hatched NW–SE = mud minnows (Umbra).

seemed to him most crucial for the question of the North Atlantic bridge. The recent earthworm family of the Lumbricidæ is, as the figure shows, distributed from Japan to Spain, but across the ocean is found only in the eastern United States. The pearl mussel occurs at the rift zone of the continents, in Ireland and Newfoundland and in the border regions on both sides. The perch family (Percidæ) and other fresh-water fish are found in Europe and Asia, but only in the east of North America. Perhaps we ought to mention the common heather (*Calluna vulgaris*), found, other than in Europe, only in Newfoundland and bordering regions; conversely, a large number of American plants are confined on the European side to western Ireland. Though it may be that the Gulf Stream accounts for the latter, this is certainly not the case for heather. Another remarkable distribution is that of the garden snail, from southern Germany via the British Isles, Iceland and Greenland across to the American side, where, however, it is found only in Labrador, Newfoundland and the eastern United States. For this case, Ökland [116] recently constructed a map which we reproduce in Figure 28. I would like to draw special attention

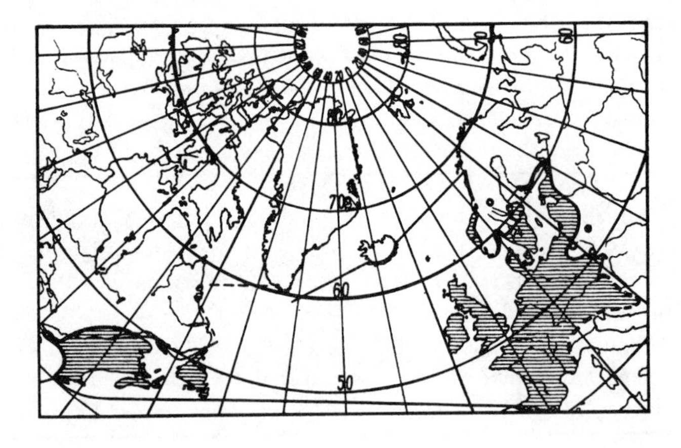


FIG. 28. More precise illustration of garden-snail distribution, according to Ökland.

here to the following consideration: Even if we neglect the fact that the theory of sunken continents is untenable on geophysical grounds, this explanation is still inferior to that given by drift theory, because it must interpolate a very long hypothetical bridge in order to connect the two small areas of distribution; with the accumulation of such cases, it becomes increasingly unlikely that the eastern and western boundaries of the distribution would have lain just on today's continents rather than on the wide continental bridge—that is, in today's ocean.

Von Ubisch [117] was right in saying: "The hypothetical bridges of the older theory as a rule extended over very considerable regions. . . . Some bridges even stretched across different climatic zones. It is therefore certain that the bridges could not have been used by all the animals on the continents that they connected, just as we do not find a completely homogeneous distribution of fauna on present-day interconnected continents, even when they extend over a homogeneous climatic zone. This is best demonstrated by Eurasia, from whose homogeneous faunal region eastern Asia is mostly separated as a special region.

"In the theory due to Wegener, matters are entirely different. According to this, the rift produced leads to the split-up of a fully uniform faunal region, provided the rift does not accidentally cut across an already existing boundary between faunas. . . .

"The consequences were bound to be specially clear when the uniform faunal region of North America and Europe was split, because the rift came relatively late and the palæontological records are correspondingly numerous. Besides, this region has been examined in a particularly detailed way, and, since the period of isolation has been relatively short, the surviving forms cannot have developed along very divergent lines.

"Actually, we could not have wished for better correlation than we find between the two regions. In the Eocene we find that almost all suborders of North American mammals occur in Europe also. The same holds for other classes. . . .

"Naturally, the close relationship between faunas on either side can also be explained by a bridge over the North Atlantic. . . . But according to what we said above, Wegener's explanation has the advantage. . . .

"Therefore, summarising our data, we may well say that the facts of zoogeography, details aside, fit excellently with Wegener's viewpoint. In many cases, drift theory is even capable of giving us simpler solutions to the situations found than those of any former theory."[24]

24 Ökland concludes, on the basis of the same material, that the sunken-continent theory, whose geophysical untenability he overlooks, is to be preferred, particularly since drift theory would lead one to expect more identities than there are. Clearly, he claims too much here; first, drift theory in no way leads one to expect *complete* identity of former flora and fauna, and second, the number of identities, both absolutely and percentage-wise, is much reduced by the incompleteness of fossil discoveries.

In a work on the Ascidiæ [130], Huus considered that the drift theory had a special advantage in that it offers not only the possibility of connections between land masses, but also proximity of habitat: "Wegener's drift theory permits a very simple interpretation of transatlantic affiliations. According to this concept, we can presuppose not only the littoral mentioned, but also a gap between the two continents which was much narrower during the Tertiary than now. It is thus possible to conceive of forms spreading right across the ocean, and the transatlantic affiliations between the central and southern parts of this ocean are easier to understand. The theory also provides a natural explanation for the close relationships between the Ascidia fauna of the West Indies and that of the Indian Ocean."

Von Ubisch [134], Hoffman [133] and recently Osterwald [120] have brought out an interesting detail about the North Atlantic region. The common spawning grounds of the American and European freshwater eels lie in the Sargasso Sea, as was discovered by J. Schmidt, and the European eel, in accordance with the greater distance to the spawning ground, goes through a much longer period of development than the American. As Osterwald rightly realised, this peculiarity is explained in a straightforward way by the gradual drift of this ocean basin plus America away from Europe; if I remember correctly, as long ago as 1922 J. Schmidt had already given me this explanation by word of mouth.[25]

[25] Both von Ubisch and Hoffmann, on the other hand, consider that these facts are contrary to drift theory and in favour of the sunken continents, but this is due to a misunderstanding: "One could at first believe that the movement of the spawning area had occurred passively, a part of the sea floor where the eels spawned during the Cretaceous–Eocene period having been dragged westwards with the American continent like a wash basin.

"But this is impossible according to Wegener's theory, because he assumes that, as the continents drift, a fresh sima surface is continuously exposed. . . ." The floor of the Sargasso Sea could not consist of newly exposed sima, but is probably identical with the floor of the ocean basin between Florida and Spain that can be seen in my map of the Eocene (Fig. 4). In reality it was most likely still smaller, because in the reconstruction the sial masses of the Azores, which should be attached to Spain and North Africa, are not taken sufficiently into account. However, it was already in existence at that time east of Florida. The crystalline covering of this basin was then, while adhering to America, displaced westwards with that continent. In a new survey paper [227], which takes into account much more of the zoogeographical literature than the paper cited here, von Ubisch recognises the given solution as a possible one, but he cloaks it in another guise—that Europe drifted eastwards, not America westwards. As the movement is relative this obviously amounts to the same thing; for if America drifted westwards

Concerning the precise moment when the rift between North America and Europe had spread to the stretch between Newfoundland and Ireland there is still a considerable divergence of opinion, as our Figure 1 shows. In any case, however, it appears to have been complete by the late Tertiary. The uncertainty in the results may be partially connected with the fact that, farther north, the bridge via Iceland and Greenland remained in existence into the Quaternary (Scharff [131] has made this seem very likely).

In this respect, investigations made by Warming and Nathorst of Greenland flora are most instructive. They show that, on the southeast coast, i.e., just on that stretch of land which extended in front of Scandinavia and northern Scotland in the Quaternary, the European elements predominate, whereas on the whole remaining coast of Greenland, including the northeast, the American influence prevails.

According to Semper [125], the Tertiary flora of Grinnell Land was, interestingly enough, more closely (63%) related to that of Spitsbergen than to that of Greenland (30%); naturally, today that situation is reversed (64 and 96%, respectively). Our reconstruction for the Eocene period gives the solution to this puzzle, the gap between Grinnell Land and Spitsbergen being narrower than that between the former and the Greenland habitats.

W. A. Jaschnov, in a work on the Crustaceæ of Novaya Zemlya [225], says that the present-day distribution of fresh-water crayfish is also best explained by drift theory: "One can state with a high degree of probability that, in hydrobiology, many problems of distribution of the lower aquatic organisms, in the northern hemisphere at any rate, can be resolved by the principles of the theory of continental drift. As an example we may mention the present scattered distribution of *Limnocalanus macrurus*, for which all forms of passive transport (i.e., by the wind and by birds) are out of the question because of the lack of any rest stages. With the presence of a connection between both continents per the Wegener theory, the range of distribution of this species was by no means a large one" (see Fig. 29).

Among other writers we mention only Handlirsch [136]. In a thorough investigation he comes to the conclusion: "Land connections

relative to Europe, then the latter drifted eastwards relative to the former. I take this opportunity to emphasise once again that the severance of South America from Africa occurred as far back as the Middle Cretaceous; for on pp. 162, 163 and 172 of the survey paper cited, faunal differences of later periods (Eocene, Miocene!) are still regarded as objections to drift theory! Cf. p. 99, note 23.

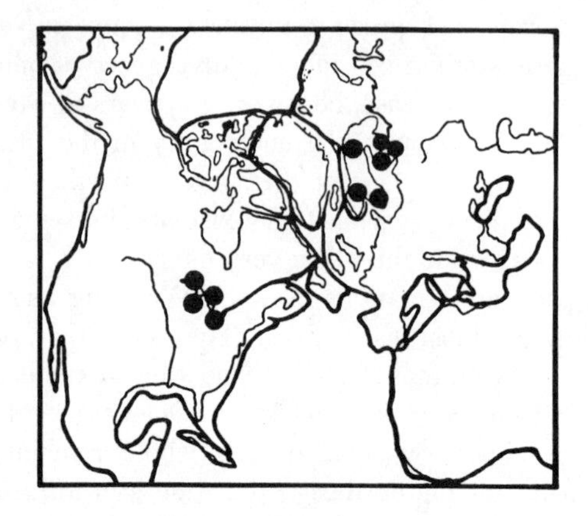

FIG. 29. Distribution of *Limnocalanus macrurus*, according to Jaschnov.

must have definitely existed between the northern part of North America and Europe, and between the former and northern East Asia as late as the Tertiary and perhaps even the Quaternary, and they must have extended over long periods of time or have occurred repeatedly. . . . However, I can find no cogent reason for assuming either direct or Antarctic Tertiary land connections between South America, Africa and Australia; I should add that no claim is thereby made that such connections did not exist even at earlier times."

Kubart [137] has made an interesting study of the flora of the islands of the mid-Atlantic ridge, which of course are considered as fragments of continents, geologically speaking. He made a statistical investigation of the indigenous types and obtained quantitative evidence, backed also by a study of the fauna, that the isolation of these islands proceeded from south to north: "Of course, these facts can be assessed as evidence not only for drift theory, but also for the existence of a large bridge-continent. In either case, the islands are viewed as the remains of these earlier processes, and even according to the land-bridge theory, the submergence of the intermediate continent bridging Africa and South America took place in a geological period earlier than that of northern Atlantis. However, according to permanence theory, the elevation of a huge Atlantis continent must be an impossibility. Therefore, this mathematical progression in the flora percentages, fully supported by the zoological data and apparently not contrary to geological findings, becomes a direct indication of the split-up of the African–European–American block, proceeding from south to north." This is exactly how drift theory sees the situation.[26]

We could cite many other authors who support the idea of the former existence of these land connections stretching across the Atlantic. However, since these connections are really no longer doubted to any appreciable extent today, it would be unnecessary to do this. As regards the evidence constituted by the distribution of earthworms, we shall return to that topic later.

The biological affiliations between the Deccan and Madagascar, alleged to have involved a sunken "Lemuria," are famous. We refer the reader to our Figure 1 and Arldt's synopsis. Diener [226], who generally supports the permanence of the large ocean basins, expresses himself on this subject as follows:

"A dry-land connection between the Indian peninsula and southern Africa via Madagascar is an inescapable feature of the Permian and Triassic periods on zoogeographical grounds. This is because in the Gondwana faunas of the East Indies, European terrestrial vertebrates . . . are mixed with such as . . . were indigenous to southern Africa. Further, the settling of Titanosaurus and a relative of Megalosaurus in Madagascar during the Upper Cretaceous must have taken place by way of India, since the Mozambique Channel had been formed earlier, during the Liassic. Not before the most recent period of the Cretaceous could the narrow, elongated island, whose ends must be sought in the Deccan and on Madagascar, have descended into the depths completely, the central section included; so that the Ethiopian mediterranean sea of Neumayr, till then a dependency of the Tethys, thenceforth was joined by a wide and unobstructed stretch of water with the Indian Ocean." Instead of the subsidence to more than 4 km depth as assumed by Diener, which is impossible isostatically over such an area, we believe that this bridge was compressed to form upland Asia. The zoogeographical difference consists in the fact that then, before the separation, the Deccan was directly alongside Madagascar. This is just where the advantages of drift theory show up, because the two areas in their present-day positions are widely different in latitude and have similar climates and support similar

[26] Naturally, Kubart is right when he says that the older concept of sunken land bridges should not be *wholly* excluded. The reader will see that, quite the contrary, use is made of the idea in many places in this book, except when considering the large ocean basins.

forms of flora and fauna only because the equator runs between them. This large separation would provide a climatic puzzle for the Glossopteris flora period, but drift theory resolves it. However, we shall not consider the palæoclimatic arguments in detail until the next chapter.

Sahni [138] has carried out a superfluous investigation, using the distribution of the polar Glossopteris flora in the region of ancient Gondwanaland in an attempt to examine the superiority of drift theory over the theory of sunken continental bridges. The question must remain undecided, however, because the observational material is still too fragmentary. That the land connection between southern Africa, Madagascar, India and Australia did in fact exist is regarded in the paper cited, as it is in all publications known to me, as a research result established long ago. However, with the vast distance which now separates these portions of the globe from each other, it is quite obviously the case, in my view, that drift theory offers a much better explanation of the facts observed than does the sunken-continent theory, which is geophysically untenable; this is also stressed by many other scientists.

The land fauna of Australia has a very special interest in this respect. Wallace already detected [139] a clear subdivision of this fauna into three elements of differing antiquity, and this result has not been radically changed by the more recent investigations of Hedley, for example. The most ancient element, generally encountered in southwestern Australia, shows some interrelationship with the fauna of India and Ceylon in particular, but also with that of Madagascar and southern Africa. Here, warmth-loving animals are representative of the relationship, and so are the earthworms, who avoid frozen ground. The correlation dates back to the time when Australia was still joined to India. According to our Figure 1, this connection had already been broken in the Early Jurassic.

The second Australian faunal element is very well-known, because it includes those peculiar mammals, the marsupials and the monotremes, which are so sharply differentiated from the Sunda Islands fauna (the Wallace limit of the mammals). This element shows points of relationship with South American fauna. Apart from Australia, the Moluccas and various South Sea islands, marsupials now live mainly in South America (one species of opossum even in North America); North American and European fossil marsupials are known, but no Asian ones have been found. Even the parasites of the

Australian and South American marsupials are the same: E. Bresslau [140] stressed that, of the flatworms, three-quarters of the approximately 175 species of Geoplanidæ are to be found in these two areas.

He states: "The geographical distribution of the Trematoda and Cestoda, which naturally corresponds to that of their hosts, has so far been only rarely a subject for special investigation. Here also there are facts of great interest to zoologists to be discovered, as we learn from the genus Linstowia of the order Cestoda, found exclusively in the South American opossums (Didelphidæ) and in the Australian marsupials (Perameles) and monotremes (Echidna)." Of this correlation with South America, Wallace has said [139]: "It is important to observe here that the heat-loving reptiles provide hardly any proof of a close affiliation between the two regions, while the cold-resistant amphibia and fresh-water fish provide evidence in profusion." The same peculiarity is exhibited by all the remaining fauna, so that Wallace considered that the land connection between Australia and South America, "if it really existed, was near their southern limits, where the climate was cold." The earthworms likewise made no use of this bridge. Since the facts point to Antarctica as the bridge, lying as it does on the shortest junction path, it should occasion no surprise that the alternative proposal of some isolated authors that the "South Pacific" bridge was the link is almost universally rejected; the latter is the shortest link only on Mercator's projection. This second Australian faunal element therefore dates back to the time when Australia was still joined to South America via Antarctica, i.e., between the Lower Jurassic (when India was detached) and the Eocene (when Australia was severed from Antarctica). The present-day position of Australia no longer isolates these forms, and they are slowly gaining ground in the Sunda Archipelago, so that Wallace already had to set the limit of mammals between Bali and Lombok and further through the Macassar Strait.

The third Australian faunal element is the most recent, and emigrated from the Sundas; it has made its home in New Guinea and has already taken over the northeast of Australia. The dingo (wild dog), rodents, bats and other mammals immigrated to Australia in the post-Pleistocene. The recent earthworm genus Pheretima made an active invasion of the Sunda Islands, of the southern Asiatic littorals of the Malayan peninsula as far as China, and of Japan, displacing most of the earlier genera; it has also already completely taken over New Guinea and obtained a firm foothold on the northern tip of

Australia. All this constitutes evidence for an exchange of flora and fauna which first began during the most recent geological period.

This threefold subdivision of the Australian fauna agrees most elegantly with drift theory. One need only look at the three reconstruction maps on page 18 and the explanation will be immediately forthcoming. *These very circumstances show in the clearest possible way the great superiority of the drift theory over that of sunken bridges even in purely biological matters.* The distance between the points in South America and Australia, nearest to each other, Tierra del Fuego and Tasmania, today amounts to 80° along a great circle, as large as that between Germany and Japan; central Argentina is as far from central Australia as from Alaska, or as South Africa from the North Pole. *Can one really be expected to believe that a mere land bridge is enough in this case to ensure the interchange of biological forms?* And how strange that Australia had no exchange of forms with the incomparably closer Sunda Islands, compared with which it is like something so foreign as to have come from another world! One cannot deny that, in this case, our view that the former separation of Australia from South America was a fraction of what it is today and that, on the other hand, it was once separated from the Sundas by a broad ocean basin over prolonged periods, does justice to the peculiar nature of the Australian animal kingdom in a way quite different from that of the theory of sunken continents, which is geophysically impossible anyway. In fact, I believe that the Australian fauna will provide the most important material that biology can contribute to the overall problem of continental drift. I very much hope that a specialist will soon be found to undertake a comprehensive study with this viewpoint as his basis.

On the question of former land bridges to New Zealand, there still appears to be no complete clarity of opinion. We have already mentioned (p. 89) that a large portion of the islands was first converted to dry land by Jurassic folding processes. At that period, New Zealand probably still formed, for the most part, a continental shelf of Australia, which, situated on the advancing side of the displacement, underwent the folding process. On the southern side, New Zealand was connected with western Antarctica and thence with Patagonia. Von Ihering [122] writes: "At the time of the Upper Cretaceous and at the onset of the Lower Tertiary, the way was clear for migration of marine animals from Chile to Patagonia and vice versa, and also to Graham Land and other parts of Antarctica as far as New Zealand." The terrestrial flora of that period in New Zealand was, according to

Marshall [141], not a forerunner of today's, but there were oaks and beeches there which presumably had come from Patagonia by way of western Antarctica, the same route which the shallow-water animals had taken. At that time there could not have been, therefore, any direct land connection between Australia and New Zealand. In the course of the Tertiary, however, such a bridge must obviously have existed, at least for a limited period, so that the present-day flora could have immigrated. According to Bröndsted [142], investigations of the sponges also reveals that the islands did at any rate have a prehistoric shallow-water link with Australia.

A work of Meyrick's on the Microlepidoptera [143] is of special interest for the question of New Zealand land connections. Apart from interesting affiliations between Africa and South America which fully substantiate the results briefly discussed above, he finds that New Zealand completely lacks a genus (Machimia) represented in both South America and Australia by many species; on the other hand, the genus Crambus occurs in New Zealand (with 40 endemic species) and has developed many forms in South America also, while in Australia there are only two species. In other words, in the first case there appears to be a connection between South America and Australia while New Zealand is out of the picture, and in the second case South America seems to be connected with New Zealand and Australia almost wholly excluded.

This, along with the facts given above, shows only that two separate migration routes led away from South America: one towards New Zealand, probably via western Antarctica, and the other towards Australia, probably via eastern Antarctica. Although New Zealand was then much closer to Australia, it seems that they were really joined for a short time only, if at all. A precise clarification of this process is naturally very much hampered by our scanty knowledge of Antarctica.

In view of what we know of the matter, the Pacific basin must have existed as such from very ancient geological times. Admittedly a number of authors have assumed the contrary; one example is Haug, who wants to explain the islands there as remains of a vast "sunken" continent; another is Arldt, who believed that the affiliations between South America and Australia should be explained by a trans-Pacific land bridge following the southern parallels of latitude—yet a glance at the globe shows at once that the route from South America to Australia crosses Antarctica. Von Ihering also assumed the existence of a Pacific continent, but the reasoning is quite untenable, as Simroth [144] demonstrated along with other matters, some time ago, and as von Ubisch [149] recently stressed. Burckhardt, too, believed in a South Pacific continent, extending from the west coast of South America westwards; however, his reason for this was just one single geological observation which, in fact, can be explained in another way altogether. At all events, this hypothesis, too, is rejected by Simroth [144], Andrée [145], Diener, Soergel and others, and even Arldt, one of its few adherents, has to concede that this land bridge is the one which has the least to recommend it [146]. Our assumption of the permanence of the Pacific Ocean since at least the Carboniferous is in accord with the overwhelming majority of investigators.

Biologically speaking, the greater age of the Pacific compared with the Atlantic shows up very clearly. Von Ubisch writes: "In the Pacific, we find many ancient forms such as Nautilus, Trigonia and the eared seal. These forms are not found in the Atlantic." Colosi [118] stressed that the Atlantic fauna, like that of the Red Sea, is distinguished by the fact that it invariably shows affiliations only with adjacent regions, while in the Pacific a characteristic feature is the existence of scattered affiliations with far-distant regions. The latter is the mark of areas where biological forms settled in ancient times, while the former is an indication of recent settlement.

Svedelius [155], in a study of the discontinuous geographic distribution of some tropical and subtropical marine algæ, has recently stated that, while the material is insufficient to test the validity of drift theory, "it should be noted nevertheless that my investigation shows how the majority of the older genera of algæ clearly have their chief distribution in the Indian Pacific Ocean, whence they migrated to the Atlantic. Only in one or two cases does the migration appear to have proceeded in the opposite direction. Therefore the algæ flora of the Atlantic should perhaps be regarded as more recent than that of the Indian–Pacific Ocean. This does not contradict Wegener's theory, which states that the Atlantic is much younger than the Indian–Pacific Ocean."

Drift theory holds that the Pacific islands and their submarine foundations are marginal chains detached from the continental blocks, gradually lagging behind in the east during the general movement of the earth's crust over the mantle in a predominantly westerly direction (cf. Chapter 8). Without going into details, their original site should therefore be sought on the Asiatic side of the ocean, to which they must, in any case, have been situated much closer in the geological period we are considering than they are today.

The biological phenomena appear to support this idea. According to Griesebach [147] and Drude [148], the Hawaiian Islands possess a flora which is most closely related, not to North America—their nearest neighbour, from which the present-day air streams and ocean currents come—but to the Old World. Skottsberg states that the island of Juan Fernández has almost no botanical connections with the nearby coast of Chile, but has with Tierra del Fuego, Antarctica, New Zealand and the other Pacific islands. However, it should be emphasised here that the biological phenomena on islands are generally harder to interpret than those on larger areas of land.

In conclusion, there are still some recent works that should be mentioned, of particular importance as the first thoroughgoing specialist treatises which take drift theory into consideration. A beginning was made by Irmscher in his large-scale investigation of plant distribution and the development of the continents [150], which appeared in 1922. This book examines present-day and former distribution of flowering plants back to Cretaceous times with a completeness unattained hitherto; it is illustrated by a large number of maps. We cannot here discuss the details of this unusually copious material.[27] The book ends as follows:

"The results justify our conclusion that three groups of factors operating in close conjunction have produced this present-day pattern of flowering-plant distribution:

"1. Shift of the poles as the cause of plant migration and intermingling of floras.

"2. Drift of large-scale blocks, resulting in changes in the overall configuration.

"3. Active spreading and further development of the plant stock."

[27] Von Ihering [122] disagrees with Irmscher because the latter gives rather different dates for part of a series of fossil plants found in South America and Antarctica from those given by the men who first investigated them. First of all, Irmscher's view does not, as von Ihering believes, constitute the arbitrary expression of a preconceived idea, but is based on expert knowledge. But apart from that, the revised dates differ so little from the original in almost all cases that it would be better to describe this as making them more accurate, not correcting them. In the meantime, Köppen and Wegener [151] have shown that, in the majority of these cases even the original dating agrees completely with drift theory and with the polar wanderings which were deduced with its help.

It is no accident that here polar wandering is mentioned first and continental drift second, because the period under consideration only extends from the Cretaceous onwards; and the nearer we approach present times, the more closely the world configuration resembles that of today, and the less continental drift can be detected in the phenomena of plant distribution. It is therefore quite natural that the large pole displacements of the Tertiary and Quaternary were foremost in making their mark on the distribution of plants. It is all the more significant, therefore, that the result supports drift theory in spite of its secondary importance. Irmscher says: "We have found that for many reasons permanence theory is inadequate to account for the facts of plant distribution and its prerequisities. However, in comparing our findings with Wegener's drift theory, it appeared that the particular features of zone structure and the prerequisites of plant distribution agree in a startling fashion with the fates of the great land masses as postulated by Wegener, the former being directly reflected in the latter.

"What permanence theory can never explain, the riddle of the Australian flora, now for the first time finds a wholly satisfactory solution. Wegener's assumption of a shift in position of the continents during the Mesozoic era is the only one that can provide the key to the otherwise incomprehensible fact that the extra-tropical forms of Australia reveal no close relations to the Asiatic ones in the way that the present-day geographical position would properly require, especially since no shift of the poles has affected this region with a disturbing influence. The former position postulated for Australia also offers the key to the problem of how this ancient flora has remained so undisturbed precisely here right up to today, how it has retained such diversity of forms and how it was able to develop further. The northward drift of Australia after its detachment from Antarctica was in fact a period of thorough isolation for this continent." One can see that the plant kingdom of Australia presents just the same pattern as does the animal kingdom!

"In the course of our investigation, we never felt the need to postulate the existence of a former Pacific continent."

Irmscher is on the right path, as one can see, in comparing drift theory, not with the theory of sunken continental bridges, quite untenable on geophysical grounds, but with the theory of permanence. Nevertheless, he gave consideration to the sunken-continents theory, but had to reject it on purely botanical evidence:

"The . . . above-mentioned North American fossil Wilcox flora, discovered in the southeastern United States (Texas to Florida), is, according to Berry's basic work on the subject, most closely related to the Alum Bay flora of southern England, also dated in the Eocene. If we now draw the equator line round the globe in accordance with the position of the poles in the Eocene demanded by Wegener, then in Europe this line would run roughly through the Mediterranean region, and England would be a bare 15° from the equator; in Asia the line would run through Indochina or thereabouts. If we assume the permanence of the present-day positions of the continents, it would mean that, for America, the equator would run through a line joining Colombia and Ecuador, from which the Wilcox flora region is over 30° off. That gives rise to difficulties in assigning the two flora even approximately the same latitude, although they require the same climate; this is because the Wilcox flora comes to lie much farther north than the southern English. However, if we follow Wegener's idea and displace America to lie alongside Europe and Africa, at one stroke we have the two flora in the same latitude, and their demands for similar climatic conditions are immediately satisfied. Here we have a case where only drift theory enables us to resolve the contradictions without exception, while the bridge theory does indeed explain the presence of similar flora on the now-separated continental blocks, but cannot provide for similarity of climate. Permanence theory must be rejected as entirely inadequate to handle this question.

"What we have demonstrated for these two flora here, holds also for the habitats of several genera that occur in the tropics. Here again the reconstruction is possible over the full range only if America is moved into zone 2 (Europe and Africa), since, with the present-day configuration of the continents, the equator in zone 1 (America) would lie too far to the south. We have already drawn attention to this difficulty and were able to find a way to remove it only by displacing the American continent. So here, for the first time, it has been shown how superior drift theory is to the theory of land bridges as seen from the standpoint of biogeography."

These last considerations of Irmscher's already lead us to the problems of palæoclimatology, but we do not want to deal with them before the next chapter.

A continuation of this important work of Irmscher's is provided by Studt's dissertation on the present-day and former distribution of conifers and the history of their regional configuration [152]; this was preceded by a shorter dissertation by Koch on the same subject [153]. Although these two authors do not agree on various botanical questions, they come to the same result so far as drift theory is concerned. Koch states: "The recent and the fossil conifer areas are fully in accord with the pole-wandering and drift theory, and only it will explain the matter satisfactorily." He adds: "For we now understand why, amongst other things, closely related species of Araucaria occur in two different areas of the globe which are widely separated by ocean; why species of Podocarpus have their habitats not only in New Zealand, Australia and Tasmania, but also in southern Africa, southern Brazil and Chile; and why Microcachrys and *Fitzroya archeri* are found in Tasmania and their corresponding forms Saxogothæa and *Fitzroya patagonica* are also found in Chile."

Likewise Studt writes: "Both the recent and the fossil distribution pattern of conifers can be explained in the simplest and least self-contradictory way by Wegener's drift theory. The extensive affiliation between the North American and European Cretaceous flora demands a continuous land connection and reduced separation between the continents. The same is true of the similarity of composition of Jurassic flora, often extending to the species themselves, which is found in regions now widely separated from each other in spite of restricted possibilities for dissemination. Only drift theory does justice to the two requirements of connection and proximity." Studt mentions that, assuming drift theory, the zonal distribution of conifers conforms far more accurately to the climatic zones, and is therefore better understood, than if one presupposes the present-day position of the continents to have been the same in the distant past.

In conclusion, we would like to make brief reference to one more source, the important work of Michaelsen on the geographical distribution of earthworms [154]. This appears to me to contain particularly strong corroboration of drift theory, since earthworms cannot tolerate sea water or frozen ground and can be transported only with difficulty, except by man.

Michaelsen shows that the attempt to explain earthworm distribution by permanence theory leads to great difficulties, while drift theory explains it "in a quite remarkable way." To illustrate the point, he makes use of two sketch maps, reproduced in Figures 30 and 31. The outlines on the maps show the earlier configuration of the continental blocks, on which the *present-day* earthworm genera are entered (no fossils are known). In regard to transatlantic affiliations,

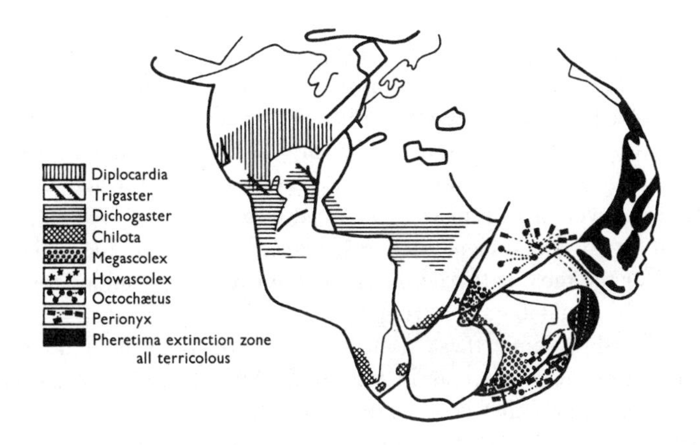


FIG. 30. Present-day distribution of some earthworm genera of the family Megascolecina, superimposed on the pre-Jurassic reconstruction based on drift theory (according to Michaelsen).

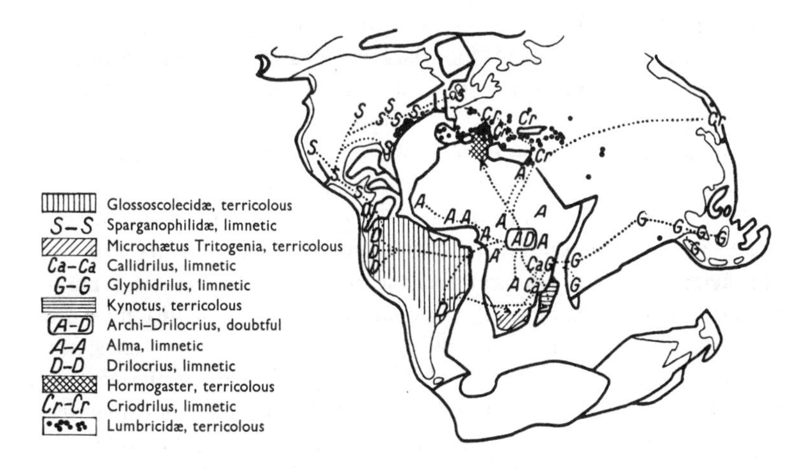


FIG. 31. Present-day distribution of the earthworm family Lumbricina, superimposed on the drift-theory reconstruction map for the Eocene (according to Michaelsen).

Michaelsen says: "I have already set forth in detail, and clarified by a tabular summary, the many interrelationships which span the Atlantic, namely those which concern five terricolous and three limnetic forms. This accumulation of regular, approximately parallel correlations makes it highly probable that direct, i.e., transatlantic connections are involved here. Wegener's theory explains these connections immediately. If one imagines that the American continent, which his theory claims broke away from Europe and Africa and drifted westwards, were now replaced and fitted into the Europe-Africa complex, the two widely separated regions either side of the Atlantic would, for the most part, run together to form a single unified region. The result would be an extremely simple distribution system. . . ." In the North Atlantic these transatlantic correlations hold between recent forms also, whereas in the South Atlantic, only the ancient forms are related; this is again in agreement with the fact that the Atlantic was opened up from the south towards the north.

After discussing the complicated correlations in India, Australia and New Zealand, which our figures show clearly, Michaelsen goes on:

"The Wegener theory of continental drift offers a remarkably simple explanation for these various transoceanic correlations of the oligochætous fauna of India. Referring to Wegener's sketchmap of the supposed approximate configuration of the continents during the Carboniferous [Fig. 30, eastern half], the first thing we observe is that India at its fullest extent (before the Himalaya was folded) reached as far as Madagascar and that its western side, the present-day Howascolex habitat (Kurg and Mysore), was joined directly with Madagascar, the second habitat of Howascolex: this explains quite simply the transoceanic correlation that holds for the western habitat of India. We can also see that the Australia–New Zealand–New Guinea block, connected on the south with the Antarctic block, extends with its northern end (New Guinea) into the angular stretch of sea (subsequently the Bay of Bengal) between India and Indochina plus the Malayan block. It is to be assumed that in still earlier times this Australian block lay with its western margin alongside the eastern edge of India.[28] This would have allowed the formation of the simple

[28] Nothing prevents us supposing that this connection was still extant in the Carboniferous and perhaps much later than that. The gap in my map of the Carboniferous only means that, so far, I know of no basis for a land link here, since this very portion of the eastern littoral of India in its elongated form lies in upland Asia as a fold system, and its congruence with the edge of the Australian block cannot be examined.

and uninterrupted lines of distribution from the south of India via Ceylon to the southernmost part of western Australia, etc. (Megascolex), and from northern India via New Guinea to New Zealand (Octochætus, Pseudisolabis) or to northern Queensland, New Zealand and southern Australia (Perionyx). It should be noted that New Guinea is a genuine member of this northern line of distribution. After the Australian block had become detached from the Antarctic, it was pushed northeastwards and its head (New Guinea), which projected to the northwest, was thrust into the Malayan block. . . . As a result of this catastrophic process, the New Guinea rammer, now in the most intimate contact with the Malayan block, was overrun by the most recent Megascolecida genus, Pheretima, which, with its great powers of self-dissemination, had in the meantime established predominance on the Malayan block; this genus divested New Guinea of its older oligochætous fauna (Octochætus, Perionyx, etc.). Thus, by the elimination of New Guinea, the gap in the line of distribution from northern India to New Zealand was enlarged to such an extent that any explanation involving a former direct land connection appeared almost impossible. By the time of this Pheretima catastrophe, New Zealand must already have been severed from New Guinea, and the Australian block also could hardly have still remained in prolonged direct connection with New Guinea, but presumably must have been cut off by a narrow stretch of shallow sea; for at the very most only a single species of Pheretima (*P. queenslandica*, apparently endemic to northern Queensland) was able to reach the Australian continent. Moreover, the separation of New Zealand from Australia, at least by shallows, must have occurred fairly early, because the former shows only slight connections with the latter. . . . Probably it was the central sections of New Zealand that first broke away in arcuate form from the Australian block, the southern end remaining at first still joined to Tasmania and the northern end to New Guinea. Then the southern end broke away from Tasmania, the northern end becoming detached from New Guinea only much later. . . . A somewhat longer-lasting, possibly isthmus-like land connection had probably formed between southern Queensland and North Island, New Zealand, by way of New Caledonia and Norfolk Island, enabling Megascolex to migrate. The route via New Guinea seems to me inadmissible because Megascolex is a typical southern Australian form. . . ."

In his conclusion Michaelsen states:

"I believe the results of my investigation may be formulated by saying that the distribution of the Oligochæta in no case contradicts Wegener's theory of continental drift but, on the contrary, should be regarded as a powerful argument in its favour; if final proof of the theory were forthcoming from elsewhere, the distribution could, in many details, be used for further elaboration of the idea. . . .[29]

"It may be said in conclusion that Wegener's sketch maps, which I used to make the distribution charts reproduced above, and on which I based my exposition, were drawn up without reference to the distribution of the Oligochæta. Only after I had informed Wegener of the remarkable way in which this distribution fits in with his theory of former land connections, did he take account of the individual facts of the distribution in the demonstration of his theory; this was done in the second, revised edition of his book on continental drift. I mention this because it appears to me to strengthen the support which the distribution of the Oligochæta lends to his theory."

[29] Michaelsen emphasises many times that the earthworm distribution points to the periodic existence of a land bridge across the Bering Strait; he believes, quite wrongly, that I dissent from this. This has never been the case. The misunderstanding may go back to an erroneous assertion of Diener's [108]: "Displacing North America in the direction of Europe ruptures its connection with the Asiatic continental block at the Bering Strait"—a mistake clearly due to reading a map employing Mercator's projection. Its untenability is immediately apparent if one takes a globe and notes that the movement of North America relative to Europe was essentially a rotation centred roughly on Alaska. (Distance between shelf margins of Newfoundland and Ireland = 2400 km; between northeastern Greenland and Spitsbergen = a few hundred km, or perhaps zero!) The same assertion has recently been repeated by Schuchert [163]; but he, too, makes a false reconstruction by rotating North America about the North Pole instead of about Alaska, a totally unfounded procedure. The previously mentioned vote-chart drawn up by Arldt concerning the existence of land bridges, including the one over the Bering Strait, shows that here there was a land connection probably as far back as the Permian and Jurassic, but certainly from the Eocene into the Quaternary. The present-day separation by the shallow shelf of the Bering Sea is therefore a very recent phenomenon.

CHAPTER 7

# *Palæoclimatic Arguments*

SINCE the last previous edition of this book appeared, the problem of the climates of the geological past was systematically examined by W. Köppen and myself [151]. The scope of our book was hardly inferior to that of the present work. Although our book was in essence a collection of geological and palæontological material, subjects in which the geophysicist and climatologist are beset by difficulties and the danger of errors that the specialist can avoid, we felt nevertheless that such an investigation was justified, since palæoclimatology can thrive only as a unification of these sciences, and the literature in this field which has appeared so far shows all too clearly that the meteorological and climatological basis for it is inadequate. Our detailed exposition will be referred to extensively in the present chapter.

However, this chapter is not going to be a review of the overall contents of our book. The task of the latter was to unravel the problems of geological climatology; continental drift is only one of many causes of variation in climate, and not even the most important factor for the more recent periods. The only question we are dealing with here is how far prehistoric climates furnish criteria for the validity of drift theory; fossil evidence of climate is therefore cited only where relevant to this. Thus the question of the origin of the Quaternary glaciation, for example, is virtually excluded, because in this period the relative configuration of continents was already so like that of the present day that but few palæoclimatic criteria are available for assessing drift theory.

The opposite is the case for the older geological periods, however, which show the most remarkable evidence for the fact that drift theory is inescapable. The number of authors who have endorsed the theory on just these grounds is by no means small.

In order to form a correct opinion, two things are necessary here: a knowledge of the present-day system of climates and its effects on the inorganic and organic worlds; and a knowledge of, and correct interpretation of, fossil evidence for climate. Both branches of research are in their infancy and leave many questions still unanswered today; all the more reason, however, to take notice of the results they have obtained so far.

The present-day climatic system as is well known, was treated by Köppen and illustrated by a map of the climates of the world [156]. This map, still not sufficiently detailed for many other purposes, is really too comprehensive for our task, since the fossil evidence allows only a very rough estimation of the climatic conditions. For this reason, we have replaced this map in our book by the simplified chart (Fig. 32) of the present-day primary isotherms and arid zones, which

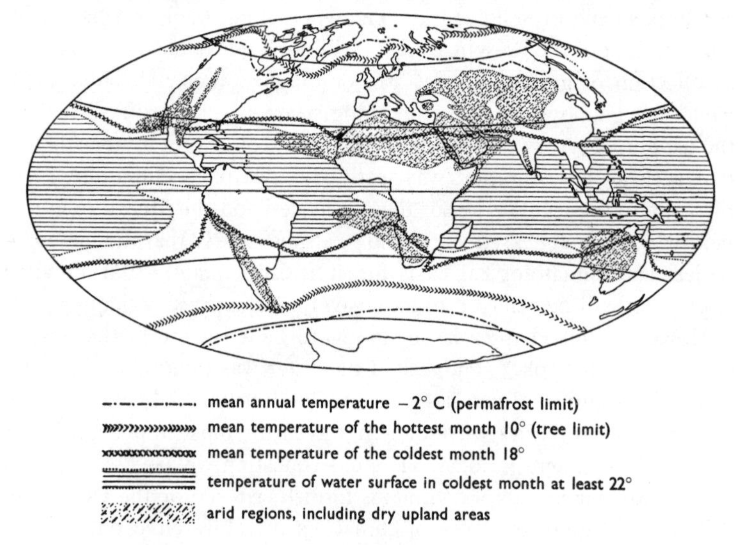


FIG. 32. Chief present-day isotherms (at sea level) and arid regions.

contains all the essentials we need. We can distinguish an equatorial rain belt with thunderstorms, which encircles the whole globe without any gap; next to this, in the high-pressure belts of the Horse Latitudes with descending air currents, are the arid regions, which are regularly interrupted on the eastern margins of the continents by the monsoon regions, but which extend far over the sea on the west coasts and push towards the poles in the heartlands of the large continents. Then follow the northern and southern rain belts of the temperate latitudes with cyclonic rainfall; beyond these lie the more or less glaciated polar caps. The belt of warm sea is entirely enclosed between the two parallels of about 28 or 30° north and south latitude. All isotherms show the predominance of a zonal arrangement of climates, but there are characteristic deviations from this caused by the distribution of land and sea: the 10°-isotherm of the hottest month, well known to coincide astonishingly closely with the tree limit, lies at higher latitudes over land than over the sea, since the land has greater annual variations than the sea. The mean annual temperature of −2° C, the approximate limit of permafrost, follows a different line. Where it runs in higher latitudes than the tree limit, it represents the same climate as that produced by inland ice (Greenland, Antarctica); where it runs in lower latitudes (as in Siberia) we find forest on a frozen soil. All inland ice is confined to latitudes of greater than 60°.

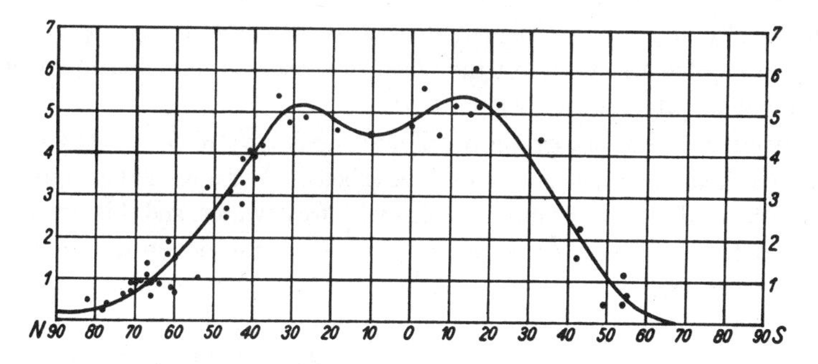


FIG. 33. Present-day altitude of the snow limit in the different latitudes (heights in km).

To supplement this, Figure 33 gives the altitude of the snow line in various latitudes, according to Paschinger [157] and Köppen [158]. It reaches its greatest altitude of over 5 km in the Horse Latitudes. The figure holds for individual mountains or for chains. In the case of extensive upland areas the snow limit lies much higher.

The geological and biological effects of this system of climates are extremely numerous. We would like now to discuss them together with the fossil evidence of climate so far available.

Perhaps the most important climatic evidence, though somewhat hazardous in nature, is the traces which former inland ice sheets have left behind. Since the crucial requirement for the formation of inland ice is a low summer temperature, which is lacking in the heartland of the large continents because of the large variations in annual temperature there, the polar climate is not always detectable as traces left by inland ice. Conversely, however, where we do find such traces, there is no doubt that we are dealing with the products of a polar climate. The most common feature is boulder clay, a name used appropriately to denote the ungraded mixture of the finest and coarsest material that distinguishes moraines. Boulder clay of earlier periods is generally cemented into solid rocks, tillites. We can, or believe we can, characterise these as Algonkian, Cambrian, Devonian, Carboniferous, Permian, Miocene, Pliocene and Quaternary. Unfortunately, it is just these most frequent signs of former inland-ice sheets which are sometimes virtually indistinguishable from other, "pseudo-glacial" conglomerates which come from ordinary detritus formation. In the latter, even polishing of the rock and scratching occur, simulating striated rock pavements, but in reality slickensides. Generally, it is usual to regard the rock as certainly glacial only if one has been able still to detect the polished surface of the outcrop under the boulder clay of the ground moraine.

Another important group of climatic signs is formed by the coals, which should be regarded as fossil peat beds. For a water basin to be transformed to peat, it must be filled with fresh water, and this process can only take place in the rain belts of the earth, not in arid zones. Coal therefore indicates a wet climate, which may be an equatorial rain belt, one in the temperate latitudes or a subtropical wet climate in the monsoon areas on the eastern margins of continents. Today, peat forms in many bogs in equatorial areas, but also in the subtropics if wet, and in the temperate latitudes, where the Quaternary and post-Quaternary peat bogs of northern Europe, amongst others, have been known the longest. Therefore, from the mere presence of coal beds one cannot derive any clue to the temperature; the character of the flora whose remains are found in the coal layers and neighbouring beds must be called upon to provide this information. A small pointer here, which should not, however, be overestimated, is given by the thickness of the coal beds, inasmuch as the luxuriant and uninterrupted growth of vegetation of the tropics can produce peat beds of greater thickness than can the slower-growing plants of the temperate zones, other things being equal.

A particularly important group of pointers to the type of climate is the products of arid regions, especially salt, gypsum and desert sandstone. Rock salt is produced by the evaporation of sea water. In most cases, this is a matter of large inundations (transgressions) of dry land, which are wholly, or to a large extent, cut off from the open sea by movements of the floor. In rainy climates, these transgressions are increasingly desalted by dilution, as in the Baltic. However, in dry climates, where evaporation exceeds precipitation, if the transgression is completely cut off, the area involved becomes continuously smaller by drying out, and the salt solution becomes increasingly concentrated; finally, salt precipitates out: gypsum is the first constituent to do this, then rock salt, and lastly potash salts, readily deliquescent. The gypsum deposits therefore usually occupy the largest area; interspersed among them one finds rock salt layers and only rarely, in restricted zones, potash salt. Even larger areas are covered by shifting sand dunes of former deserts, now hardened into sandstone; these are marked by lack of vegetation and animal life. They do not constitute the reliable evidence for dry climates that salt and gypsum do, since sands and dunes, though to a lesser extent, also occur in wet climates as shore formations, like those of present-day northern Germany; they even occur at the front of the edge of inland ice, like the *Sandur* of Iceland. A clue, though somewhat indefinite, to the temperature situation is provided by the colour of these sandstones; in the tropics and subtropics a red colour predominates in the soil formation; brown and yellow in the temperate and high latitudes. Shore sands are, of course, white in the tropical zones, too.

For marine deposits, a law which applies is that thick lime beds can be laid down only in the warm waters of the tropics and subtropics. Although bacterial activity appears to play some part, the probable reason is simply that cold polar water can dissolve much greater amounts of lime and is therefore unsaturated; whereas warm tropical water, which can hold much less lime in solution, is saturated or supersaturated (cf. the precipitation of boiler scale, or "fur" in kettles). Obviously related to this also is the generally much greater lime precipitation of tropical organisms, primarily the corals and calcareous algæ, but also mussels and snails. In polar climates, deposition of massive limestone beds appears to be generally impossible, just as limestone disappears from the ocean sediments proper, due to the low temperature of the deep sea.

In addition to these inorganic signs of climate we still have those of the plant and animal kingdoms, with which, of course, greater caution must be exercised, because the organisms are very adaptable. From a single find, therefore, one can rarely draw any conclusions; but usable results can always be obtained if one keeps in mind the overall geographical distribution of the plant or animal world of a given period. By comparing contemporary flora from different parts of the world, one can generally decide with great certainty which of the two lived in the warmer climate and which in the colder, even though the absolute value of temperature can be estimated only for the more recent geological formations, where the plants are already similar to those of today; for the older flora, absolute temperature remains mostly undetermined. Lack of annual rings in trees implies a tropical climate; prominent rings mean a temperate one, despite the exceptions to this rule, which are not at all rare. Where trees were lofty, we may probably assume for prehistoric times also that the temperature of the warmest month exceeded 10° C.

The animal kingdom also provides many climatic criteria. Reptiles, which do not generate their own body heat, will die in a cold-winter climate from the freeze-up, which renders them defenceless. They can therefore live in such climates only if they are sufficiently small to take cover easily, like our lizards and grass snakes. If, as in polar regions, there is, moreover, no summer heat, their eggs will not hatch out in the sun, so that they generally cannot find conditions there that make survival possible. Therefore, where reptiles have developed with a rich abundance of forms, it can be concluded that the climate was tropical, or at least subtropical. Herbivores generally provide a criterion for the vegetation and hence for the amount of rainfall; fast runners, such as horses, antelopes and ostriches, indicate a steppe climate, since their anatomy is adapted to master the problems of large open spaces. Climbers, like monkeys or sloths, are at home in forests.

It is not possible to discuss the details of all such climate evidence here; what has been said should, however, be enough to give a rough picture of how one generally arrives at conclusions about prehistoric climate.

The prodigious number of facts which can be utilised in this way as fossil evidence of climate now shows surprisingly that in prehistoric times most parts of the earth had very different climates from those of today. It is known for example, that Europe had a subtropical to tropical climate throughout most of the world's history. As late as the beginning of the Tertiary, central Europe had the climate of the equatorial rain belt; then, in the middle of this period, followed the formation of large salt deposits, i.e., a dry climate developed. Towards the end of the Tertiary the climate was rather like that of today, and then followed the Quaternary glaciation, i.e., a polar climate at least for northern Europe.

A particularly noticeable example of large climatic changes is the region of the North Pole, especially the case of Spitsbergen, the best known. This area is separated from Europe by only a shallow sea and thus forms a part of the great Eurasian continental block. Today Spitsbergen has a harsh polar climate and lies under inland ice; but in the lower Tertiary (when central Europe lay in the equatorial rain belt) forests grew there with a wider range of species than is to be found in central Europe today. There were not only pines, firs and yews, but also lime trees, beech, poplar, elm, oak, maple, ivy, sloe, hazel, may, guelder-rose, ash and even such warmth-loving vegetation as water-lily, walnut, swamp cypress (Taxodium), mighty sequoias, plane trees, chestnut, gingko, magnolia and the grapevine! It is obvious, therefore, that the prevailing climate at Spitsbergen must have been about like that of present-day France, which implies that the mean annual temperature must have been about 20°C higher than today's. If we go back still further in world history, we find indications of even greater warmth: in the Jurassic and Lower Cretaceous, Spitsbergen had sago palms (now found only in the tropics), gingko (today found as a single species in China and southern Japan), tree ferns and other such plants. Furthermore, as far back as the Carboniferous, we find there not only thick beds of gypsum, indicating a subtropical, dry climate, but also a flora which has an equally subtropical character.

This enormous climatic shift—in Europe from tropical to temperate, in Spitsbergen from subtropical to polar—immediately suggests a shift in the position of the poles and the equator, and thus the whole zonal system of climates. In fact, this suggestion is inescapably confirmed by the *equally large, but exactly reversed climatic change experienced by South Africa* (80° S of Europe, 110° S of Spitsbergen) *in the same period*: in the Carboniferous it was buried under an inland ice sheet, i.e., it then had a polar climate, today a subtropical.

These fully authenticated facts permit of no other explanation than that of polar wandering.[30] We can make yet another test for this. If the meridian through Spitsbergen and South Africa passed through the

[30] On the concept of polar wandering see Chapter 8.

greatest climatic change, then the simultaneous climatic change in the two meridians 90° E and 90° W of this must have been nil or quite insignificant; and this is in fact the case. The Sunda archipelago, 90° E of Africa, definitely already had the same tropical climate in the Lower Tertiary that it has today; this is shown by the unaltered preservation of many ancient plants and animals, such as the sago palm or the tapir, and recently Carboniferous plants have also been found there of the same type as those found in Europe, which are held to be tropical by the best experts. The northern part of South America was also in the same situation; the tapir has been preserved there also, amongst other forms, but it is found only as a fossil in North America, Europe and Asia, and not at all in Africa. Of course, the constancy of climate in the northern part of South America is less complete than for the Sunda Islands; this, as we will see, is a consequence of the drift of continents: South America at one time was not 90° W of the Spitsbergen–South Africa meridian, but much closer.

After what has been said, it is not surprising that those who have attempted to fathom the system of prehistoric climates have referred to polar wandering time and again from the very outset of the investigation. In his *Ideen zur Philosophie der Geschichte der Menschheit*, Herder already hinted at such an explanation of prehistoric climates. Thereafter it was advocated in greater or less detail by many writers, namely: Evans (1876), Taylor (1885), Löffelholz von Colberg (1886), Oldham (1886), Neumayr (1887), Nathorst (1888), Hansen (1890), Semper (1896), Davis (1896), Reibisch (1901), Kreichgauer (1902), Golfier (1903), Simroth (1907), Walther (1908), Yokoyama (1911), Dacqué (1915), E. Kayser (1918), Eckardt (1921), Kossmat (1921), Stephan Richarz (1926) and many others. Arldt [159] has collated this literature up to 1918, but since then the number of authors in favour of polar wandering has snowballed.

Formerly, this theory encountered general opposition within the narrow circle of professional geologists, and up to the time when the works of Neumayr and Nathorst appeared the majority of geologists completely rejected the idea of polar wandering. After these works appeared, however, the picture was changed insofar as adherents of the concept among geologists increased, though slowly. Today, the overhelming majority of geologists supports the view formulated in E. Kayser's *Lehrbuch der Geologie*, which is that the assumption of a considerable polar displacement during the Tertiary is "difficult to avoid"—although some opponents still rejected this view a few years ago with a severity which is hard to understand.

Meanwhile, however cogent the reasons for belief in polar wandering during the earth's history, it is undeniable that all early attempts to determine the position of the poles and the equator at all points of time have led to absurdities so grotesque that it is no surprise that the concept was suspected to be erroneous. Such systematic attempts, mostly undertaken by outsiders, have therefore never attained recognition; examples of this are works by Löffelholz von Colberg [4], Reibisch [161] and Simroth [162], Kreichgauer [5] and Jacobitti [164]. One of them, Reibisch, unfortunately forced his quite correct concepts of the subject from the Cretaceous onwards into the amazing straitjacket of a strict "pendulation" of the poles in a "cycle of oscillation"; this is probably incorrect as a physical law of gyroscopic spin, is in any case unfounded and, what is more, leads to many contradictions of observed facts. Simroth has collected a comprehensive body of biological facts in order to prove the pendulation theory. This material contains good evidence for pole wandering, but has no power to convince one of the strict regularity of the pendulation as claimed. Obviously, it is more correct to proceed by induction, that is, simply to deduce the position of the poles from the fossil evidence for climate without preconceived ideas about the result. This was the method of Kreichgauer in particular; his book was clearly written, but he based it on an insufficiently substantiated dogma about the configuration of the mountains, alongside the real evidence for climate. Almost all these efforts give about the same result for the more recent periods as did Köppen and I: The North Pole at the onset of the Tertiary was situated near the Aleutians; from there it wandered towards Greenland, where it is to be found at the beginning of the Quaternary.[31] For these periods, no great internal discrepancies exist. However, it is otherwise for the periods before the Cretaceous. Here, not only to the views of the above-mentioned authors diverge widely, but all these reconstructions lead to hopeless contradictions, since they presuppose the immutability of relative continental positions

[31] This position of the poles in the Lower Quaternary has recently been reconfirmed in a striking manner by a number of biological facts adduced by von Ihering [122] from South America; Köppen has made reference to this [127]. Admittedly von Ihering himself would prefer to explain these facts by an alteration of the pattern of ocean currents, the present-day position of the poles being assumed. In my opinion, this is inadmissible, but we cannot discuss the matter in detail here since it is outside the scope of this book.

as self-evident. In a quite characteristic manner, these contradictions are such that they absolutely prohibit *every conceivable position of the poles*.

However, if one starts out from the standpoint of drift theory, and if one maps the fossil evidence for climates on a chart developed for the relevant period by means of the theory, these contradictions completely vanish, and all the climatic evidence arranges itself to form the pattern of climatic zones which is familiar to us today: two arid belts, between which a humid zone runs on a great circle round the globe and which, with the humid belt, include all evidence for tropical heat; outside these, on both sides, two humid belts; where evidence of polar climate is found, the centre thereof is 90° on a great circle from the centremost humid belt and about 60 great-circle degrees from the nearest dry one.

We now consider the Carboniferous as the oldest period for which maps based on drift theory have so far been drawn up. Here we encounter at once the most difficult problem of palæoclimatology to date, the traces of Permo-Carboniferous glaciation.

All the present-day southern continents (and the Deccan) were glaciated at the end of the Carboniferous and the beginning of the Permian; however, apart from the Deccan, no continent of the northern hemisphere was glaciated in this period.

It is in South Africa that these inland ice traces have been most accurately studied; where Molengraaff in 1898 first found the ice-polished bed rock under the old moraine, and removed any lingering doubts about the moraine-like nature of the "Dwyka conglomerate" there [165]. Later investigations, among which particular stress is laid upon those of du Toit [166], give us a very detailed picture of this glaciation. In many places one can read the direction of movement of the ice from scratches on the polished rock; from this, a series of centres of glaciation can be determined, from which the ice spread outwards; and one's attention is drawn to slight time differences in the main activity of each of these centres, which correspond all told to a shift in the greatest ice thickness from the (present-day) west towards the east. From the 33rd parallel southwards in South Africa, the boulder clay lies concordantly on marine deposits and appears to be a direct continuation thereof; the only way to interpret this is to say that the inland ice terminated here as a floating "barrier," as in present-day Antarctica, whereby the ground moraine, which melted out at the lower edge, lay over the earlier marine sediment as its natural continuation. The snow line must therefore have been at sea level here. The very extent of this South African glaciation, almost equal to that of present-day Greenland, indicates that we are dealing with a genuine inland ice sheet, not just a montane glacier phenomenon.

However, exactly the same moraine deposits are also found on the Falkland Islands, in Argentina and southern Brazil, in India and in western, central and eastern Australia. In all these areas, the glacial interpretation of the hardened boulder clay is fully ensured by the complete similarity of the whole bed series. They all lay under inland ice sheets, as in South Africa. In South America and Australia several superimposed boulder-clay beds have been found with interposed interglacial deposits—just as in the Quaternary glacial and interglacial periods in northern Europe. For example, in the central part of eastern Australia (New South Wales) there are two moraines, separated by coal-bearing interglacial beds; therefore, the land here was twice overrun by inland ice, but in the interim there were freshwater lakes over the moraine landscape which turned to bogs. South of this area, in Victoria, there was only one glacial layer; north of it, in Queensland, none at all. The southern-most part of eastern Australia was therefore continuously buried under ice in this period, the central area was invaded by ice only twice, and the north remained entirely free. Thus, exactly the same pattern begins to reveal itself here as the one we have long been familiar with for the Quaternary ice age of Europe and North America. In the latter area, the alternation of glacial and interglacial eras can be attributed to periodic variations in the orbit and axial inclination of the earth and hence in the value of the solar constant. It must be accepted as certain that such variations have occurred during the whole history of the planet. Noticeable effects could, however, only be left behind in times when inland ice lay over the polar caps. All these details show clearly that the Permo-Carboniferous glaciation of the southern continents is a genuine inland-ice phenomenon.

However, these traces of the Permo-Carboniferous ice age are now widely separated and involve almost half the surface of the globe!

Now consider Figure 34. Even if we place the South Pole at the most favourable point conceivable in the centre of these traces of glaciation, i.e., at about 50° S, 45° E, then, as the equator associated with this positioning of the poles will show, the inland-ice traces farthest from the poles in Brazil, India and eastern Australia lie at a geographical latitude of not quite 10°; a polar climate would thus

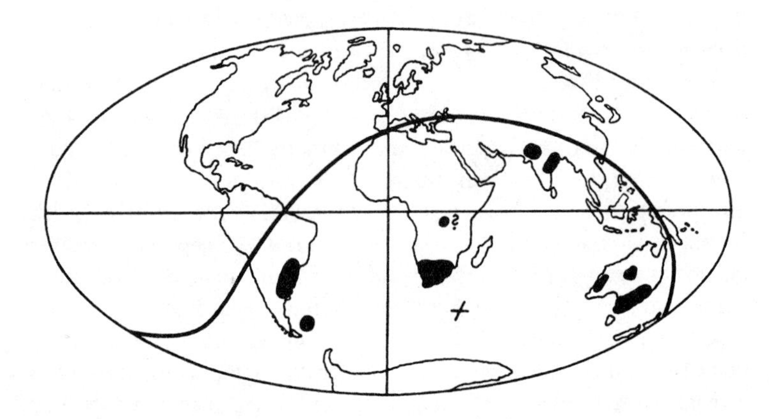

FIG. 34. Traces of the Permo-Carboniferous inland glaciation on present-day continents. The cross indicates the position of the South Pole most suitable for the explanation; the broad curve is the associated equator.

have prevailed almost up to the equator. The other hemisphere would have, as we anticipate, only traces of tropical and subtropical warmth as far as Spitsbergen. It is not necessary to stress that this result is ludicrous. The attempt to explain these ice traces climatically was made by Koken [167] as far back as 1907, when the South American discoveries could still be regarded as uncertain, and it amounted virtually to a *reductio ad absurdum*; his conclusion—that apparently no other assumption is possible except that all these traces of the glaciation were made at considerable elevations above sea level—must be ruled out because even highlands of such an extension do not produce inland ice in the tropical zones; also, the observed facts prove just the reverse—the snow limit here was reduced to sea level. Since then, in fact, no new attempt has been made to explain the phenomena by climatology.

The result is that these traces constitute a striking defect in the hypothesis of continental immobility. What would we think of the drift theory if any part of its large collection of material led to such contradictory conclusions? The permanence of position of the continental blocks has hitherto been treated as an *a priori* truth, requiring no proof. Yet in fact it is merely a hypothesis which must be checked against the observed facts. And I very much doubt whether geologists are in a position to adduce more cogent proof of any of their results than the Permo-Carboniferous ice traces provide for the invalidity of permanence theory.

We shall refrain here from citing the literature in support of our statements. The obvious needs no backing by outside opinion, and the wilfully blind cannot be helped by any means.

As far as we are concerned, it is not now a question of whether the continental blocks have moved; doubt is no longer possible. It is rather a question of whether they have moved in accordance with the particular assumptions of drift theory.

First of all, we should not overlook the fact that in a number of other places in the Permo-Carboniferous beds conglomerates have been found which also have so far been considered by geologists as glacial and the situation of which fits in less well, and in part quite poorly, with the special ideas of drift theory.

For example, from central Africa, Permo-Carboniferous (and also Triassic) conglomerates have been reported [216] which have so far been identified with the South African Dwyka conglomerate and have been interpreted as ground moraines of an inland ice sheet. Permo-Carboniferous traces of glaciation in the Congo region could be tied in with drift theory if absolutely necessary (Triassic traces, however, could not very well), but in my opinion this would require assumptions which are unlikely on climatological grounds. But how certain is this glacial interpretation? We have already mentioned that deceptively similar "pseudo-glacial" conglomerates with polished sections can also arise in quite different climates (especially in dry climates) and have been proved to do so. The polished rock beneath the supposed moraine has not yet been found in the Congo, so that the only signs so far are those also typical of pseudo-glacial conglomerates. Besides, the bed series there is known only from small fragments—even the classification as Permo-Carboniferous is unreliable—so that no one can say that the glacial interpretation is confirmed by the identity of the whole series of beds. The little we know of these beds seems, on the contrary, to point to an essentially different formation and thus to an origin under quite other climatic conditions. In no way, therefore, can the glacial interpretation be regarded as reliable. We also have the direct objection that it is believed that the northern limit of inland ice can be determined in southern Africa. It is hard to believe that another, separate ice cap was also present in central Africa at the same time. One is therefore justified for now in disregarding the conglomerates of central Africa as evidence of climate. I think it probable that their pseudo-glacial nature will later be revealed.

This is even more likely to prove the case with the Permo-Carboniferous conglomerates found by Koert in Togo, which are also claimed as glacial in the investigations made to date, not very thorough as yet. In my opinion, however, they were very probably formed in a dry climate.

There is another series of conglomerates claimed as glacial in North America and Europe that it is quite impossible to fit into the overall picture given by drift theory, which is otherwise so consistent. For example, Hobson believed that there were traces of glaciation from the Carboniferous period in the Ruhr basin; Tschernischev thought he could detect some in the Urals, from the Upper Carboniferous.

In the same way, W. Dawson found supposed traces of glaciation in Nova Scotia in 1872, confirmed by A. P. Coleman in 1925; S. Weidman (1923) found some in the Arbuckle and Wichita ranges in Oklahoma; J. B. Woodworth (1921) in the "Caney shales" of Oklahoma; Udden in the Permian beds of west Texas; Süssmilch and David mention also the "fountain" conglomerates of Colorado. These instances are today regarded as pseudo-glacial by the overwhelming majority of geologists; in this they are surely correct, because a glacial interpretation would contradict all the other climatic evidence from these areas, of which there is a considerable amount. Van Waterschoot van der Gracht [210] writes on the matter as follows:

"We must be very cautious about 'tillites.' I consider it as unproven that any of the Permo-Carboniferous conglomerates of Texas, Kansas, Oklahoma and particularly Colorado can be regarded as glacial in origin. Anyone familiar with cloudbursts of the kind that occur in deserts or on the borders of arid zones would not be at all surprised that ungraded, mostly clastic and partly angular material is deposited in considerable thicknesses by the floodwaters produced by such downfalls. These floods are very violent, though short-lived. The rivers consist generally more of mud than water, and the mixture has so high a specific gravity that not only can it transport unbelievably large boulders, but any elutriation is prevented. No ice is needed to explain this phenomenon. We can see the same processes nowadays in all deserts, those of the American West included.

"Single large blocks in otherwise fine marine deposits need not have been transported by floating ice. Large trees can produce the same result when they carry large rocks, held in the roots, onto the surface of a lake.

"Even polished and gouged rocks need not have been glacial, except when the scratches are very frequent and the rocks consist of very dense, hard material. Such rocks, with an astonishing resemblance to the glacial boulders and erratics, from the Permian conglomerates of northwestern Europe, and with clear indications of 'glacial' character, are now regarded merely as fragments scratched in landslides. In 1909 I myself once made the mistake of describing one of these European conglomerates as tillite."

In addition to the cases mentioned, we have a particularly remarkable phenomenon in a Permo-Carboniferous conglomerate discovered near Boston, in the United States. This has been named "Squantum tillite" and has been interpreted as a hardened moraine by all investigators to date, especially Sayles [168], who gave the most accurate description. These deposits cover an area almost as large as the Vatnajökull in Iceland. The conglomerate contains polished rocks that are considered to be ice-gouged boulder detritus, and round this region hardened clay beds are found similar to the Quaternary and post-Quaternary varves studied by de Geer in Sweden. However, all these phenomena could be pseudo-glacial. The polished rock under this supposed moraine has not been found so far.

As I recently emphasised [217], there are very serious doubts about the glacial interpretation of this Squantum tillite from the climatological standpoint, and they are quite independent of drift theory. All other evidence for the climate of North America in the Permo-Carboniferous (a very large amount of data) shows unambiguously that the West of the United States had a hot-desert climate throughout this period, while the East still lay in the equatorial rain belt in the Carboniferous, but in the Permian was also included in the hot-desert region. Further details are given below about these signs of climate, in which the principal rôle is played by salt and gypsum deposits and coral reefs. Now, our Figure 33 shows that in climates where such deposits are produced the snow limit runs at the highest altitude it ever attains on the earth's surface. At the period in question, the line must have run at over 5 km above sea level in the region of the United States. It seems completely impossible, therefore, that in the midst of these depositions a mass of ice comparable with the Vatnajökull can have lain there; or that icebergs floated in the same sea where coral reefs formed, as many believe. Such things would be physically impossible, because the climate cannot have been simultaneously hot and cold. Nor can one get anywhere with the idea that these glacial formations arose at great altitudes. I regard it as very probable, therefore, that the Squantum tillite will also turn out to be pseudo-glacial, as so many other conglomerates already have.

It should be noted here that these climatological misgivings about the glacial nature of Squantum tillite derive from those deposits of the North American block which are its neighbours in both time and space; that is, the objections have nothing at all to do with drift theory, and demand an explanation without reference to it.

It is therefore illogical to see in Squantum tillite an objection to drift theory. However matters stand with Squantum tillite, it is obvious that we must judge our theory by the large number of reliable and mutually consistent pieces of evidence, not by the one anomaly which has already proved misleading in so many cases.

I have discussed the pseudo-glacial phenomena of the Permo-Carboniferous in some detail here because I still seem to stand alone in my protests against the glacial interpretation of Squantum tillite, and therefore had to give detailed reasons for my objection.[32] We now turn to the examination of how the *dependable* evidence of climate during the Carboniferous and the Permian becomes consistently arranged when drift theory is assumed!

The most important factors are entered on the maps of Figures 35 and 36. Genuine traces of glaciation are indicated by the letter *E*. As one can see, all the regions then glaciated are centred around South Africa and are contained in a cap of about 30° radius over the earth's surface. The contemporary signs of polar climate are thus confined to the same area as in the present-day climatic system. No better corroboration of our theory could be desired.[33]

[32] Only van Waterschoot van der Gracht appears to share my doubts [210].

[33] A fallacious objection has been put forward here that since the glaciation of the southern continents did not occur quite simultaneously, one could make do with the present-day position of the continents if one assumed simply polar wandering (very extensive and rapid, to be sure!). But the first glaciation of Australia took place as far back as the Carboniferous, at the same time as that of South America and Southern Africa, and in view of the enormous migration of the South Pole involved, the North Pole would have had to cross Mexico—but a hot desert climate prevailed there. All the other evidence of climate distributed over the whole surface of the earth most decidedly contradicts the idea of such extensive wandering of the poles.

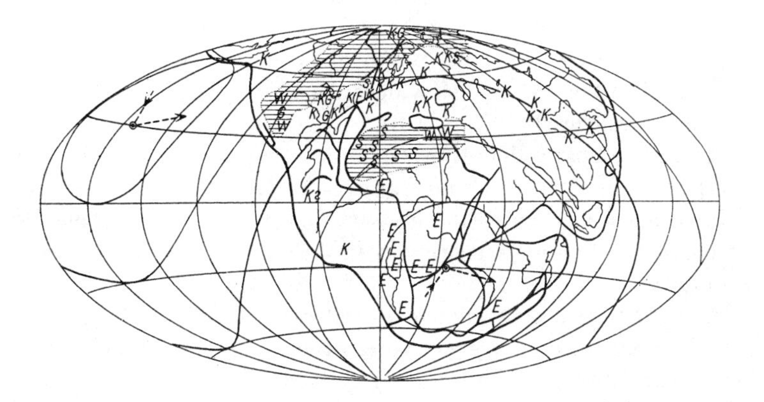

FIG. 35. Ice, bog and desert in the Carboniferous.
$E$ = ice traces; $K$ = coal; $S$ = salt; $G$ = gypsum; $W$ = desert sandstone; hatched = arid zones (according to Köppen and Wegener).

How does it come about that no indications of inland ice at the North Pole cap are found to match the large number at the South Pole? The explanation is that the North Pole was in the Pacific at a point far removed from all continents.

The centre of the glaciated region in the figures is the South Pole; the equator proper to this construction, the parallels 30° and 60° N

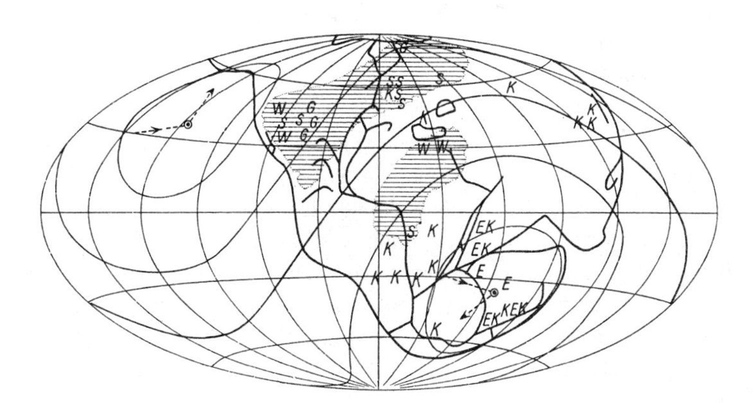

FIG. 36. Ice, bog and desert in the Permian.
$E$ = ice traces; $K$ = coal; $S$ = salt; $G$ = gypsum; $W$ = desert sandstone; hatched = arid zones (according to Köppen and Wegener).

and S and the North Pole are drawn in also. These curves naturally appear extremely distorted in the projection used; the equator, really a great circle on the globe, is shown as the curved line somewhat heavier than the others. How does the rest of the climatic evidence fit into the picture?

In our reconstruction (not on the present-day globe!) the vast Carboniferous coal belt, traversing North America, Europe, Asia Minor and China, forms the great circle whose associated pole lies in the centre of the glaciated region; the belt coincides with our equator as drawn.

Coal signifies a rainy climate, as stated previously. Any rain belt which, as here, *forms a great circle* round the globe must obviously be equatorial only. If we can establish in addition, as in this case, that the belt is 90° *from the centre of a large region of inland ice*, we are all the more justified in concluding that it occupied an equatorial position.

It is important to realise that this conclusion is quite inescapable, regardless of whether we start from drift theory or not. The European coal fields of the Carboniferous period today lie exactly 80° north of the traces of inland ice of the same period in southern Africa, traces which have been so thoroughly examined and are so reliable; it is in southern Africa that we have proof that the snow line extended down to sea level, as it does today only in Antarctica. In view of the Alpine compression in the Tertiary, the separation must have been 10°–15° more in the Carboniferous than it is today, but otherwise the position of Europe relative to South Africa cannot have been essentially different. There cannot be the least doubt therefore that the European coal fields in the Carboniferous were exactly 90° removed from the centre of a large inland ice region at the time of their formation; this holds regardless of what assumptions one makes about the positions of the other continents at that time. At 90° from the pole one can only be at the equator. Spitsbergen, too, is still part of the European continental block and therefore must have had essentially the same position relative to Europe as today. Its large Carboniferous gypsum beds betoken a subtropical dry climate and thus show that the northern zone of such climates still extended 30° N of the European coal beds at that time.

*The conclusion that the European Carboniferous coal beds were formed in the equatorial rain belt is therefore unavoidable,* without any reference to drift theory.

This evidence is so compelling that by comparison all other criteria must take a back seat. Nevertheless, it is naturally justifiable to ask whether the nature of the plant remains found in the European coal beds of the Carboniferous and in nearby beds agrees with this result. In the judgement of the best authority on European Carboniferous flora, H. Potonié, it does. His investigation of this point [169] remains today the most thorough and the best; purely on botanical grounds he concluded that the European Carboniferous coal beds were fossil peat bogs of the same character as tropical low-lying swamps.

The reasons adduced by Potonié for this belief are naturally not of a conclusive nature, for it is very difficult to assess the climatic character of so old a flora. The uncertainty involved has been strongly emphasised by his opponents, of which there are quite a number amongst present-day phytopalæontologists. It is nevertheless remarkable that (so far as I know) they are unable to weaken Potonié's case by finding another more probable climatic interpretation for the characteristics of the flora he cites, or by identifying other characteristics of the flora not mentioned by him which would indicate a different climate. It is always objections of a general nature that are brought up by Potonié's opponents. Precisely for this reason, since his botanical reasoning apparently remains quite unimpugned, it is not without interest to become acquainted with it. There are mainly six characteristics of the flora which imply a tropical origin:

1. So far as can be judged by the reproductive organs of the fossil ferns, they were related to families which have a tropical habitat today. Among other features, the relationship between many Carboniferous ferns and the present-day Marattiaceæ is worth mention.

2. In the Carboniferous flora, tree ferns and climbing or twining ferns are much in the foreground. There is a general predominance of tree-like growths even in groups which today are mostly herbaceous.

3. Many Carboniferous ferns, such as the tree fern Pecopteris, have aphlebias, i.e., imperfectly formed (irregularly serrated) pinnæ at the attachment area of the side stems; these are strikingly different from the other regularly formed pinnæ, and are already full-grown when the young, normal pinnæ are still coiled up. Such aphlebias are seen today only on tropical ferns.

4. A considerable number of Carboniferous ferns have fronds of a size that occurs only in the tropics. There are fronds of several square metres in area.

5. Annual (growth) rings are completely lacking in the trunks of

European Carboniferous trees. Growth was therefore interrupted neither by periodic drought nor periodic cold. We can now add this: On the other hand, Permo-Carboniferous trees with clearly marked annual rings have been found in the Falkland Islands and in Australia, both of which lay in the high southern latitudes, as Figures 35 and 36 show.

6. Stem-borne blossoms ("cauliflory") have been determined as characteristic of "Calamariaceæ and lepidophytes, in the latter case certain Lepidodendraceæ (of the genus Ulodendron, established solely by those large marks on the stem which correspond to stem-borne blossoms) and Sigillariaceæ. . . . Today, trees whose blossoms sprout from the side of old wood (trunks, branches) are almost entirely confined to tropical rain forests. . . . Perhaps it is the intense struggle for sunlight—the result of the dense tropical vegetation-cover—that expresses itself in the fact that the foliage, which requires this light, is often found only at the treetop, while the reproductive organs occur on those parts of the plant less accessible to light, where, in any case, they in no way obstruct the vigorous vital functions of the foliage."

These botanical conclusions may be regarded as uncertain, as stated above, but two things can be said with certainty: This flora lived neither in the cold polar climate nor in the temperate climate which prevails today in its places of discovery; only a tropical or subtropical climate can be involved here. Secondly, all the indications fit very well with our result, which was obtained by quite other and much more reliable methods, namely, that these coal beds were formed in the equatorial rain belt.

Potonié's opponents mostly take the view that we are dealing here with a subtropical, not a tropical climate. I do not know whether it is still the case, but at one time the reason given was that no peat bogs exist in the present-day equatorial rain belt, and cannot do so since, it was claimed, peat is not formed above a certain temperature due to more rapid decomposition of plant sections in the heat. The simplest way to dispose of this line of thought is to point out that, in recent times, peat bogs have been found almost everywhere in the present-day equatorial rain belt, in particular in Sumatra, Ceylon, Lake Tanganyika and British Guiana. Many others are to be found in the swamp areas of the Congo and the Amazon Rivers; these are as yet unknown directly, but their existence is made probable by the tea-coloured "black water" of many of the rivers there. This objection is therefore nothing but a mistake caused by the inaccessibility of tropical swampland and our resultant lack of knowledge of them so far. The formation of peat bogs was, of course, specially facilitated at the Carboniferous equator by the bed-rock movements of the great Carboniferous folding processes, which began at the same time; these processes disturbed the natural water run-off, and particularly extensive swamps were produced.

As a further reason for the assumption of a subtropical climate, the fact has been adduced that tree ferns, frequently found in the Carboniferous coal beds, today occur less in the tropics than in the subtropics, where they grow on well-watered mountain slopes. On the one hand, however, this is not a conclusive reason, since even today tree ferns do occur, though relatively seldom, in the peat bogs of the equatorial rain belt, and it is not unlikely that they have been partly replaced there today only by more recent, better adapted forms which were not yet extant in the Carboniferous and thus could not then compete with them. On the other hand, the comparison with present-day subtropical regions is not altogether valid insofar as they are dry until they reach the monsoon areas on the eastern margins of the continents; this means that a marsh belt as extensive as the principal Carboniferous coal fields cannot be accommodated in the subtropics of today, for climatological reasons. Coal belts can only correspond to equatorial or cool-temperature climates; but tree ferns are out of the question in the latter type.

Finally, even if Potonié's interpretation has been regarded with scepticism by many authors just because he is said to have misinterpreted the Tertiary lignites with regard to climatology,[34] we may well choose to disregard this; after all, the conclusion that who errs once, errs every time, is most certainly even less reliable than Potonié's arguments in favour of the tropical nature of the European coal beds of the Carboniferous.

This whole dispute over the tropical or subtropical nature of these coal beds is being conducted on the basis of arguments of an inconclu-

[34] Without embroiling myself in the phytopalæontological dispute, I would like to take this opportunity to mention that central Europe, according to the total evidence relating to climate, was *undoubtedly* still in the equatorial rain belt in the Lower Tertiary, in the subtropical (partly arid) climate zone in the Middle Tertiary and had a climate about like today's in the Upper Tertiary. The Tertiary coal beds of central Europe must therefore have been formed under very different climatic conditions, depending on their age. Note here, too, that climate can be determined far more reliably by the total fossil evidence of Europe's climate at that time than by the single body of evidence provided by the coal flora.

sive nature, which is not surprising in the case of so ancient a flora. All the same, I must repeat that the site of these beds a quarter of a great circle distant from the centre of an undoubtedly polar inland ice sheet is an *utterly compelling reason* for believing them to have originated in a climate of equatorial rains; this is quite independent of the problem of continental drift, as I stressed before.

Drift theory only completes the proof by taking into account the sections of this huge coal belt which lie outside Europe; if drift theory is ignored, the present-day position of this belt leads to contradictions.

The similarity of flora and hence of the climatic conditions of origin is nowadays generally acknowledged in the case of the great Carboniferous coal beds of North America, Europe, Asia Minor and China. Since the European part must have of necessity originated in the equatorial rain belt, the same must hold for the other sections. Their present-day configuration provides a direct proof of drift theory, since it does not meet the requirement that all these sites must lie on a great circle. By way of illustration, Figure 37 gives Kreichgauer's [5] map of the world in the Carboniferous with the equator as assumed by him; we see here the pattern one is led to believe must have existed if one leaves out drift theory: for Europe, Africa and Asia, there is rough agreement with our own picture. However, the equator here does not go through the eastern United States, where it belongs by virtue of the evidence of climate, but through South America, where it cannot possibly have lain, since here, scarcely 10° away from it, the inland ice sheet was expanding. Again, naturally, the incompatibility of the position of India and Australia with their traces of glaciation is particularly striking here.

Further, the great thickness of the coal beds of the principal coal belt of the Carboniferous, which makes them so valuable, is in excellent agreement with their origin in the equatorial rain belt. Much less thick are the coal beds which were formed everywhere over the southern continents during the Permian on the ground moraines of the melted ice sheets (see Fig. 36). The associated flora, called after the herbaceous fern Glossopteris, belonged to a polar climate. We are dealing here with bogs of the southern subpolar rain belt formed in just the same way as the Quaternary and post-Quaternary peat bogs of northern Europe and North America. These coal formations and the Glossopteris flora also mean that these regions must have been joined, because they now include far too large an area for their climate of that time to have covered.

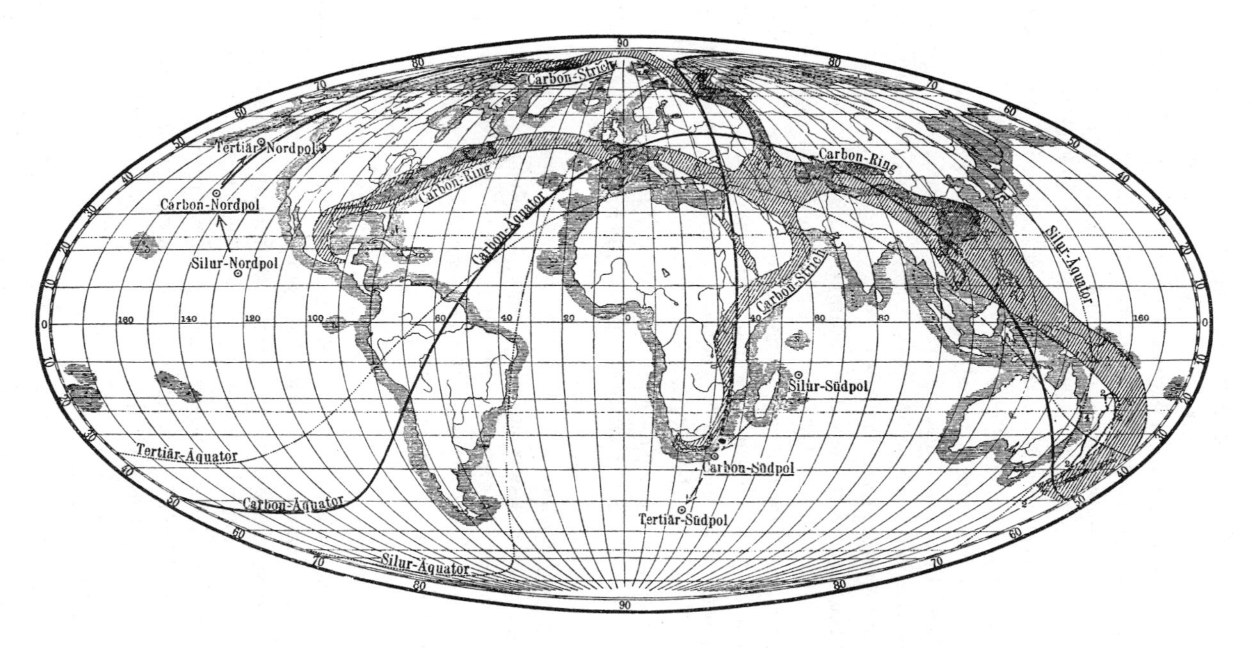


FIG. 37. Equator line and folds of the Carboniferous, according to Kreichgauer.

The other data which indicate the climates of the Carboniferous and Permian periods also corroborate the results shown in our Figures 35 and 36, where the zonal pattern is only realised if the position of the continents as conceived by drift theory is accepted.

Of the two subtropical climatic belts which contain the arid zones, the northerly one can be particularly well followed during the Carboniferous and the Permian, with regard not only to its actual existence, but also to its progress to the south in the Permian, whereby the equatorial rain belt was expelled from North America and Europe and replaced by a dry climate: During the Carboniferous, large gypsum beds were laid down on Spitsbergen and in the west of North America (*G* in Figure 35), and in the latter area the thick Permo-Carboniferous red beds are everywhere an indication of desert climate. The equatorial rain belt only lay in the east of North America. But in the Permian, all North America and Europe were desert: In the most recent Carboniferous in Newfoundland, salt already covered the last coal beds (*S* in Figures 35 and 36); in the Permian, large gypsum deposits formed in Iowa, Texas and Kansas, and salt deposits in Kansas. In Europe, crossed by the equatorial rain belt in the Carboniferous, the great salt deposits of Germany, the southern Alps and southern and eastern Russia were formed in the Permian. For Germany alone, Arldt [11] counted nine Permian salt beds, among which was the famous Stassfurt deposit. This southward displacement of the climatic zones in Europe and their simultaneous shift to the southeast in North America, together with the movement of the inland ice from southern Africa towards Australia, indicate a polar wandering from the Carboniferous to the Permian, albeit a moderate one.

The southern arid region, so far as one can deduce from the observations to date, has left traces of the Carboniferous age mainly in the Sahara, where many large-scale salt deposits originated, and also in the desert sandstone of Egypt. Admittedly, these deposits have not been examined even nearly so thoroughly as those of Europe, particularly insofar as accurate dating is concerned.

Lastly, the Carboniferous coral reefs of Europe (Ireland to Spain) and North America (Lake Michigan to the Gulf of Mexico) fit naturally into the pattern of climatic zones; so do the Richthofeniidæ (which form limestone reefs) of the Permian period in the Alps, Sicily and eastern Asia.

It is clear from the foregoing that not only the Permo-Carboniferous traces of glaciation, but also the total climatic evidence of that period falls into place with the application of drift theory and forms a climatic system which corresponds completely to that of today, provided the South Pole is displaced to southern Africa. With the present-day position of the continents, however, it is altogether impossible to combine the data into an intelligible system of climates. These observations therefore constitute one of the strongest proofs of the validity of drift theory.

The palæoclimatic evidence for drift theory would be incomplete, of course, if it held only for the Carboniferous and Permian and not for the succeeding periods. (For the earlier periods, palæoclimatic proof is not available at present, because the cartographical basis is still lacking.) However, this is not the case by any means. In the book I wrote with Köppen [151], I have treated each succeeding geological period in the same way as I have here (in abbreviated form) treated the Carboniferous and the Permian. The restricted scope of the present book forbids repeating these discussions and we must therefore refer the reader to the Köppen–Wegener book. However, the result remains unaltered: If one makes use of the reconstruction map based on drift theory as a starting point, the climatic evidence always arranges itself to form a system basically like today's, but if the present-day position of the continents is taken, contradictions result. The nearer we approach the present time, the less marked, naturally, these contradictions become, since the configuration of continents approximates more and more closely that of today, and the less convincing does this evidence become for the truth of drift theory.

For the rest, it should be noted that in the interpretation of ancient climates, polar wandering, above all for the later periods, plays the most important rôle. Polar wandering and continental drift here form, in mutual supplementation, the cardinal principle; by its use the previous confusion of disordered, apparently self-contradictory facts links up to form a pattern of a simplicity that astonishes one again and again and is extremely persuasive by virtue of its complete analogy with the present-day climatic system. However, this is primarily thanks to drift theory, for, without it, the theory of polar wandering can give at best a passably satisfactory solution for the most recent periods.

CHAPTER 8

# *Fundamentals of Continental Drift and Polar Wandering*

THE expressions "continental drift" and "polar wandering" are at times used in quite different senses in the literature which has appeared so far, and there is some confusion about their interrelationship which can only be resolved by a precise definition. Such a definition is also necessary in order to distinguish clearly the problems which these words comprise.

The assertions of drift theory relate entirely to *relative displacements of the continents*, that is, to displacements of portions of the earth's crust relative to an arbitrarily chosen portion. In particular, the reconstructions of Figure 4 (p. 19) show continental drift (displacement) relative to Africa, so that Africa has the same coordinates in all reconstructions. Africa was chosen as the reference area because this continent represents the core of the former primitive block. If consideration is confined to one portion of the earth's surface, it will be natural to locate the reference system at a limited zone of this portion, and then to present this reference zone as constant in position. The choice of zone is a matter of pure convenience. Because of the recently introduced monitoring of geographical longitude changes, the system may later be changed so that continental drift is presented everywhere as relative to the Greenwich Observatory.

In order to free oneself from arbitrary selection of a reference system, one could perhaps define *balanced continental displacements* which would be determined relative to the whole earth's surface instead of merely to a portion thereof. However, their determination would be fraught with great practical difficulties, and at the moment cannot be considered.

It is important to realise the complete arbitrariness of the African reference systems we used. When, for example, Molengraaff [228] stresses that the mid-Atlantic ridge shows that Africa drifted from there towards the east, I cannot discern any disagreement with drift theory in his statement. Relative to Africa, America and the mid-Atlantic ridge drifted westwards, the former at about twice the rate of the latter; relative to the ridge, America drifted westwards and Africa eastwards at about the same rate; relative to America, both the ridge and Africa migrated eastwards, the latter twice as fast as the former. On the basis of relative movements, all three statements are identical. But once we choose Africa as the reference system, we cannot assign a displacement to this continent, by definition. We have already pointed out that this choice of system can be the most expedient, not for the individual portions of the earth's surface, but only for the whole surface.

Continental drift as defined still says nothing about longitude changes of the pole or the substratum. I think it is important to separate these concepts from that of continental drift.

*Polar wandering* is a geological idea. Since only the uppermost layer of the crust is accessible to geologists, and since the former site of the poles can only be estimated by means of fossil evidence as to climate, which originates on the surface of the earth, we have to define polar migration as a surface phenomenon, that is, as a rotation of the system of parallels of latitude relative to the whole surface of the globe, or otherwise as a rotation of the whole surface relative to the system of parallels, which amounts to the same thing because all movements are relative. To be effective, this rotation must obviously be about an axis which differs from that of the earth's axis of spin. The question of the earth's interior — whether it remains at rest relative to the system of parallels or relative to the surface of the globe, or thirdly, which is also possible, rotates relative to both—is wholly ignored in this definition. That is necessary in order to keep things clear. Superficial polar wandering in this sense can only be detected in the remote past by the fossil evidence for climate. Geophysics cannot make any judgement about its reality or possibility.

The determination of polar wandering as defined is, of course, difficult because of the simultaneous occurrence of continental drifts. If there were no continental displacements, the positions of the poles as derived from fossil evidence of climate could be directly compared with one another, and the direction and extent of polar migrations could be found at once. But if in the interval between the two times in question, continents have drifted, we can, it is true, by means of climate evidence, find the polar sites even on the two reconstruction charts which take drift theory into account; but here arises the peculiar difficulty that we do not know at time 2 where to place the "unchanged" pole position corresponding to time 1; yet this "undisturbed" position must be known to establish the vector displacement.

One might proceed here somewhat as follows: If we imagine the map grid system at time 1 firmly engraved on the earth's surface of that period, then at time 2 the grid would be distorted because of continental displacements. If we now look for the undistorted grid which conforms most closely to the distorted version,[35] then its poles represent the "undisturbed" poles for time 2, and the comparison between their positions and the actual time–2 poles (as derived from fossil data on climate) gives the extent of polar migration between times 1 and 2.

The result would be the *absolute* value of superficial polar wandering. Because of the difficulty we mentioned, no attempt to determine it has yet been made; everyone has always been content to give the *relative* amount of wandering, i.e., relative to some arbitrarily chosen continent. Köppen and I [151] again used the continent of Africa as our base, and have thus described the polar wandering relative to Africa. If another continent were chosen as reference, naturally the polar wandering would be quite different. Only if there were no continental drift would one always find the same (i.e., the absolute value of) drift of the poles. How different relative polar migration turns out to be according to the choice of reference continent, as a result of continental drift, is illustrated by Figure 38, which shows the polar wandering since the Cretaceous, relative to Africa in one case and to South America in the other.

The observations of the International Latitude Service imply that polar wandering is taking place today. This migration also can only be related to the earth's surface. It is a milestone in the development of our knowledge of polar movements over the earth that there has been this recent success in deducing present-day progressive polar wandering, when up till then it had never been possible to do more than determine periodic cycles of change about a constant average position of the pole. In 1915, Wanach first deduced a displacement of this mean position, but he was unable to vouch for this at the time because

[35] We need not here go into the mathematical conditions involved.

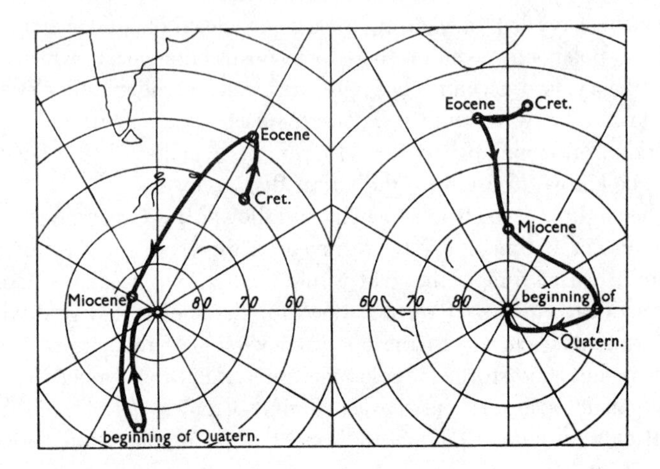


FIG. 38. Wandering of the South Pole since the Cretaceous; LEFT: relative to South America; RIGHT: relative to Africa.

the shift was so small.[36] The first quantitatively reliable evidence was presented by Lambert in 1922, and recently Wanach [208] produced a new derivation of polar wandering based on the Latitude Service's observations from 1900.0 to 1925.9. We have reproduced one of Wanach's illustrations in Figure 39, which shows the quantities involved very clearly. The total polar shift, as is well known, follows a quasi-circular path, with now a larger, now a smaller radius of curvature, as the pole of rotation (for the instantaneous axis) moves round the pole corresponding to the axis of inertia. To prevent the figure becoming too involved, Wanach has indicated only three sections of the total polar shift, namely the one with a specially short radius from 1900.0 to 1901.2, that with a very large radius from 1909.9 to 1911.1 and the small-radius section from 1924.7 to 1925.9. The pole of inertia of the earth, which always forms the centre of the figure, and is found by means of a balancing calculation, has shifted along the short oblique line in the centre of the figure. Its annual shift, the present-day annual polar wandering, amounts to 14 $\pm$ 2 cm or 140 km (1.3°) per million years, which exceeds the Mesozoic polar wandering deduced from geological data, but is less than that of

[36] By 1912 I already mentioned in *Petermanns Mitteilungen* (p. 309) that the systematic displacement of the centre of the curves described by the pole could be detected by the eye, which is extremely sensitive to symmetry of shape.

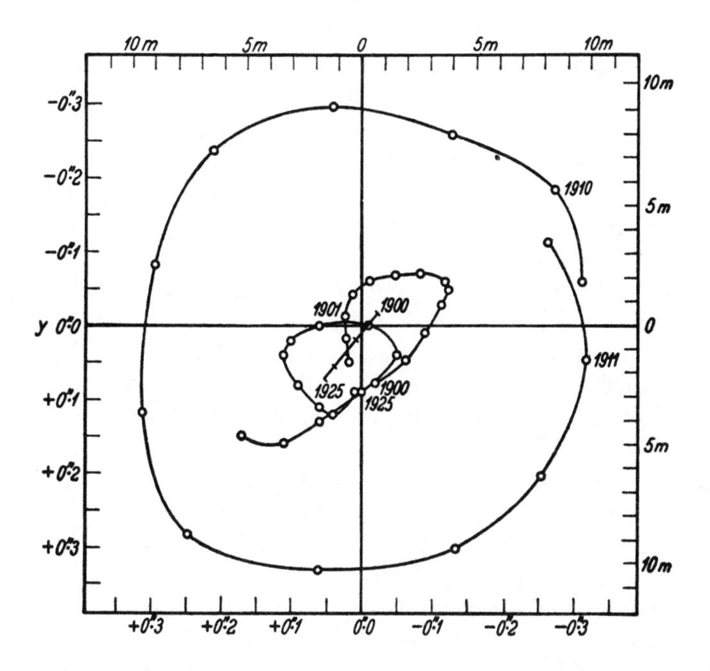


FIG. 39. Polar wandering from 1900 to 1925, with selected sections of the total polar movement (according to Wanach).

the Tertiary. If the speed and direction remained constant, the North Pole would reach the southern tip of Greenland in 23 million years.

This present polar migration does not correspond theoretically to the relative polar wandering as referred to a single continent, but rather to the absolute polar wandering as referred to the whole surface of the globe, though the two are not completely identical. This is because the latitude stations are distributed over the whole world. Nevertheless one should note that, to deduce the absolute value of polar wandering, it would, strictly speaking, be necessary to take measurements of pole latitude at all points of the earth's surface, so that the International Latitude Service can only furnish us with an approximation to the absolute amount of polar wandering. This amount could only be found accurately if the stations of the Latitude Service retained their present sites unaltered by continental drift. However, that they do shift is apparent from the situation to which Schumann [220] called attention: derivation of the pole track yields residual errors which are systematic and therefore cannot be errors of observation, although their origin is not clear at first sight.

In my opinion it is very important to define polar wandering in the way given above—as *superficial*—and so to separate from the determination of its reality the controversial question whether it arises from displacement of the crust over its substratum, or from internal shift of the axis. In the literature to date the problem has not been treated this way, and the results are confusion and perplexity. So far, polar wandering has been detected by geologists in an empirical manner (and the present-day polar shift has been deduced by geodesists from latitude determinations); many geologists dispute its possibility on theoretical grounds; and a third group of authors makes a compromise proposal, that the shifts may have arisen as a result, not of displacement of the internal axis, but only of rotations of the crust over the substratum. To escape from the confusion, it is necessary to form more rigorous concepts, and the first step towards this is to define polar wandering as superficial; such surface migrations have been detected for past geological periods and for the present time, and therefore there is no point in discussing their possibility.

By "crustal wandering" and "crustal rotation" we mean movements of the earth's crust relative to the substratum. The word "crust" implies the antithesis of the interior of the earth, so this defintition is a natural one to make. We have many types of evidence for crustal migration over the substratum, but the data allow an assessment of only the direction of displacement, not its magnitude.

First of all, we have many indications of an overall crustal rotation, in a westerly sense, and therefore proceeding about an axis which corresponds to the axis of rotation. Associated with this is the phenomenon of small blocks lagging to the east relative to large ones. Examples are the marginal island chains of eastern Asia, the West Indies, the South Shetland island arc between Cape Horn and Graham Land; also the pointed tips of the continents curved eastwards, such as the shelf areas of the Sunda archipelago and of Florida, the southern tip of Greenland and of Tierra del Fuego, the northern tip of Graham Land; further, the split-off of Ceylon, the easterly drift of Madagascar from Africa, and of New Zealand from Australia; one must also mention the Andean compression. It is true that all these phenomena come first under the classification of continental drift; but they betoken a systematic displacement of the continental blocks westwards relative to the neighbouring sima of the ocean floor, and therefore imply that the continental blocks have probably also moved westwards relative to their underlying sima. Since these indications can be traced over the whole globe, they constitute a sign of an overall westerly rotation of the crust. In fact, much use is already being made of this concept in present-day geophysics.

On the other hand, certain phenomena betoken a partial crustal wandering, in particular in the direction of the equator. This would be expected theoretically because of the existence of a force applied to the continents in a direction away from the poles. The vast Tertiary fold system from the Atlas Mountains as far as the Himalaya indicates a compression in the direction of the equator of that period, which could only have come about by virtue of a migration of the crust over its substratum.

All the foregoing indications are indirect. A more direct sign of crustal wandering over the underlying layer is provided by the gravitational field distribution. We must now go somewhat more deeply into this.

Figure 40 is a map of the gravitational anomalies in central Europe, drawn up by Kossmat [38]. The actual observed values of the acceleration of gravity were, as usual, reduced as though the whole relief pattern of the earth had been planed off to sea level and the measurements carried out at this sea level; that is, apart from the reduction to sea level, the effect of the land masses above this level was subtracted from the result. Each reduced experimental value was then compared with the normal value of gravity for the geographical latitude of the locality in question, and the difference—the gravitational anomaly—was indicated on the map. The figure shows us directly the mass deficit beneath the mountains, which partly compensate for it by the isostatic process. Kossmat states: "Here one can only arrive at the same conclusion that many geophysicists have already expressed, and Heim has also voiced, that it is not a reduction in density of packing that caused the deficit, but rather that, as a result of folding, the upper, relatively light constituents of the earth's crust were much densified, and that this bulge sank into the plastic substratum as it was being formed. The folded range did not only grow upwards, but also downwards, due to its weight: the fold upthrust has its counterpart in an even larger fold downthrust, as Heim puts it." We can therefore obtain from the map an idea of the approximate topography of the underside of the sial crust; it is beneath the Alps, where the gravitational anomaly attains its highest negative

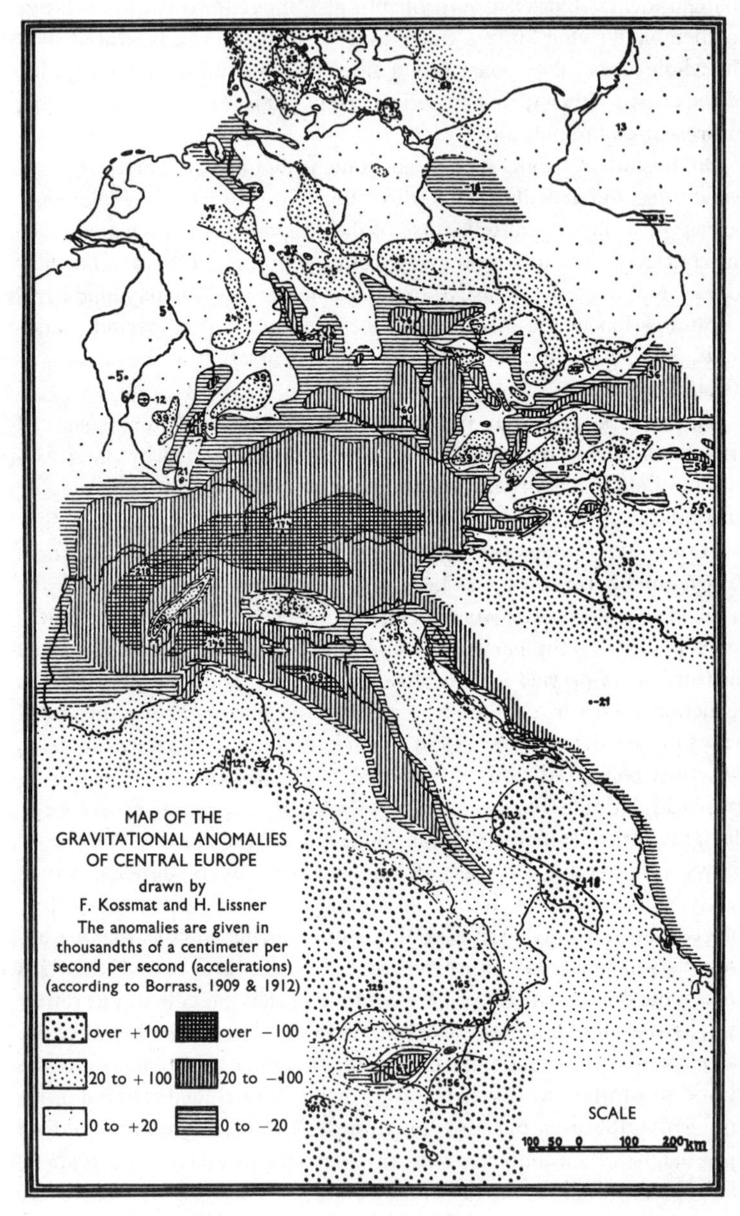


FIG. 40. Gravitational field perturbations beneath the central European mountains, according to Kossmat.

value, that the underside of the sial also sinks most deeply into the sima.

However, the important task for us is the more precise comparison of the position of this subterranean fold mass relative to that of the mountain ranges; for this, we ask the reader to refer to an atlas. It will then be easy to see that *the negative gravity anomaly is systematically displaced northeastwards.*

This striking fact betokens a subterranean bulge, tipped and carried as a whole in a more or less northeasterly direction. Now, this definitely implies a movement of the European continental block to the southwest relative to its underlying sima, during which its downward projections into the sima were held back by friction. If we had similar maps of the gravitational anomalies for the whole world, then at any rate wherever there were recent block thickenings, we could determine the direction of movement relative to the underlying sima. This would appear to be the sole direct method for determining the crustal migration. In Europe it proceeds towards the southwest, and therefore has a westerly component, which may correspond to the overall westerly rotation of the crust; and a southerly component, corresponding to crustal migration towards the equator.

We would now like to try to answer the question whether superficial polar wandering can be produced by displacement of crust over the substratum.

Clearly, this could only be a matter of an overall crustal rotation, and one whose axis is quite different from the axis of rotation of the earth. However, the observations signify that such a crustal rotation took place as a whole only in a westerly direction, i.e., about the rotational axis of the earth; one would think that any overall crustal rotation about some quite different axis would be also detectable in the configuration of the earth's surface. Observation, therefore, does not corroborate the proposed solution to the question. And what has theory to say? Theory supports both a partial, equatorwards crustal wandering and an overall westerly migration, precisely the two displacements empirically indicated: this is so by virtue of the force acting away from the poles, and the forces of the tides and of precession. Clearly, however, there is no possibility of explaining theoretically an overall crustal rotation that must take place about an axis quite different from the rotational axis of the earth. The well-meaning compromise suggestion of many authors that polar wandering can be explained by an overall crustal rotation thus lacks both empirical and theoretical support. It therefore appears very unlikely to me that the suggestion is correct. If this solution is wrong, then superficial polar wandering can only be accounted for by internal axial displacement.

The words "axial displacement" immediately suggest a shift of the axis within the medium which surrounds it over its whole length; we shall therefore use the expression in this sense only. It is still possible to differentiate between the internal axial shift in the globe and the astronomical axial shift relative to interstellar space. Initially, we wish to discuss only the first concept.

To the question whether the manifest superficial polar migration comes from an internal axial shift, one can, as will be shown, make a theoretical as well as an empirical approach. As far as the theoretical side is concerned, it has been asserted again and again by many authors that internal axial shifts of the required order of magnitude are impossible; to prove this, Lambert and Schweydar, for example, have calculated that even a displacement of Asia by 45 degrees of latitude would result in a shift of the earth's principal axis of inertia of only 1 to 2 degrees. It is obvious that these assertions and calculations by such distinguished geophysicists make a strong impression on geologists, who are in no position to test and appraise the assumptions behind the calculations. These statements have therefore led to a disgraceful state of confusion, the removal of which seems to me to be a pressing duty of the theoretical geophysicist.

The opinions of such eminent theoreticians as Lord Kelvin, Rudzki and Schiaparelli ought to make one stop and think. Lord Kelvin says [212]: "We can not only concede the possibility, but even assert it as in the highest degree probable, that the axis of maximum inertia and the rotational axis, always close to each other, could have been in ancient times far removed from their present-day geographical position, and that they could have drifted gradually by 10, 20, 30, 40 or more degrees without any perceptible sudden disturbance either of land or sea having arisen." Rudzki writes [15] in just the same vein: "If ever the palæontologists became convinced that the distribution of climatic zones in one of the past geological epochs indicates a rotational axis quite different from that of the present day, geophysicists could do nothing but accept the postulate."

Schiaparelli [211], in a little-known work, has treated the problem in rather more detail. W. Köppen [200] has given a summary of his line of thought. Schiaparelli examined the three cases of a completely solid earth, a completely fluid earth, and thirdly, one which behaved as a solid up to a certain limiting value of applied forces, but began to flow once this value was exceeded; in cases 2 and 3, unlimited axial shift was found to be possible.

How does it come about that other authors are so adamant in their rejection of internal axial shifts? The simple answer is: because they assume incorrectly that, in these processes, the equatorial bulge of the oblate earth remains unaltered in position! All denials of internal axial shift start out from this not only unfounded, but certainly inadmissible assumption.

If we make this false assumption, it is clear even without calculations that the principal axis of inertia of the earth, and hence the rotational axis also, are determined once and for all. The equatorial radius of the earth is 21 km longer than the polar. The equatorial bulge therefore represents an enormous mass extending around the earth's equator, which has a moment of inertia about the earth's axis vastly greater than those associated with the equatorial diameters of the earth. Even the largest geological variations can therefore only lead to changes in the mass distribution that are of negligible magnitude by comparison with this bulge due to flattening. If the latter remains constant, one can see even without performing any calculations that the earth's principal axis of inertia can be changed by a minimal amount only. And the rotational axis must always remain in the neighbourhood of the principal axis of inertia.

I must confess, however, that it is difficult to see how anyone can seriously assume today that the equatorial bulge could have kept its position unchanged, as though the earth were absolutely solid. The occurrence of isostatic equilibrations and relative displacement of continents are enough to show that the earth has a finite degree of fluidity, and if this is so, the equatorial bulge also must be able to re-orient itself. We need only follow along the lines of thought set out by Lambert and Schweydar: Let us assume that the inertial pole (without alteration of the bulge) has been displaced by a small amount $x$ as a result of geological processes. The pole of rotation must follow suit. The earth now rotates about an axis which is slightly different from the previous one. It must follow that the equatorial bulge re-orients itself. The earth is viscous, so the re-orientation proceeds slowly, and it is possible that it does not reach completion, but stops short before that point. We know nothing about the latter possibility. As a first approximation we must undoubtedly assume that complete re-orientation is achieved, even if the time taken is long.

Once it is achieved, however, we have again the same state as after the onset of geological change: the geological driving force acts as before and shifts the principal axis of inertia by the amount $x$ in the same direction, and the process repeats itself indefinitely. Instead of a single displacement by the amount $x$, we now have a *progressive* displacement, whose rate is set by the size of the initial displacement $x$ on the one hand, and by the viscosity of the earth on the other; it does not come to rest until the geological driving force has lost its effect. For example, if this geological cause arose from the addition of a mass $m$ somewhere in the middle latitudes, the axial shift can only cease when this mass increment has reached the equator or, rather, when the equator has reached it.

Naturally, this problem requires thoroughgoing mathematical treatment. The elementary considerations given, however, are sufficient in my opinion to show that to assume an unchanging bulge of oblateness is to make a fundamental mistake, one which leads to complete misrepresentation of the problem in question. I do not believe there is the least theoretical ground for doubting the possibility and the reality of very large, though gradual, internal axial shifts in the course of geological time. However, it would be most desirable for a theoretical attack to be made on the problem soon from some valid starting point; of course, the treatment will not be so simple as when the assumption of a solid, constant equatorial bulge is made.

However, as already mentioned, it is possible to reach a verdict by empirical methods also. Admittedly, the available methods of deciding whether surface polar wanderings are produced by axial shifts are indirect and therefore less reliable. But it is remarkable that all methods which enable a judgement to be formed have so far pointed to the reality of axial shifts.

We first recall Figure 40 and the southwesterly migration of the crust of Europe over its substratum deduced from it. The sial ridge of the European mountain ranges, held back to the northeast, was forced downwards mainly during the Tertiary; from this we may reasonably assume that the southwesterly crustal wandering of Europe had also already begun by the onset of the Tertiary. However, in the course of the Tertiary, the latitude of Europe increased by about 40°, the North Pole came that much closer to Europe, while Europe simultaneously shifted over its substratum towards the equator! That is obviously only possible if an internal axial shift took place whose amount somewhat exceeded even that calculated for the earth's surface. The only way to get around this conclusion would be to assume that the displacement of the negative gravitational anomalies towards the northeast in Europe first dated from the Quaternary, and that in the Tertiary the deficit lay *southeast* of the mountain ranges at all times. This cannot perhaps be quite ruled out, but it appears to me very unlikely.[37]

We now come to another empirical test possibility, namely the transgression cycle.

Many authors (Reibisch, Kreichgauer, Semper, Heil, Köppen *et al.*) have already discussed the fact that internal axial shifts must be tied up with systematic transgression cycles; this is because the earth is ellipsoidal and because there is a time lag while it adjusts itself to the new position of the axis, whereas the sea follows at once. Figure 41 explains this: Since the ocean follows immediately any re-orientation of the equatorial bulge, but the earth does not, then in the quadrant in front of the wandering pole increasing regression or formation of dry land prevails; in the quadrant behind, increasing transgression or inundation. Since the equatorial radius of the earth is about 21,000 m greater than the polar, then with the 60° polar wandering between the Carboniferous and the Quaternary, if it was accompanied by an equal amount of internal axial shift, Spitsbergen would have had to rise about 20 km above the surface of the sea, and central Africa would have had to sink a similar amount below, if the earth had retained its shape. Naturally, the latter cannot have been the case, because the possibility of large axial shifts depends on its re-orientational flow. However, the adjustment probably involved a lag of the order of magnitude of 100 m behind the immediate re-setting of the ocean level, and this must have shown up as transgression cycles.

[37] In his great work on the structure of the Alps [18; similarly in 215] Staub writes: "Europe and Africa drifted northwards together. From Permian times, Europe fled from Africa but the mighty colossus finally restrained little Europe in the Middle Tertiary, extruded the floor of the former ocean, which then lay between them, as a vast mountain range over Europe and thrust it further north. The amount of continental drift was . . . 50 degrees of latitude for Africa and about 35–40° for Europe." To describe the latitude shift of Europe as continental drift is a definite confusion of ideas. The result is an unfounded and, in all probability, false picture of the process, involving the two concepts: (1) Africa and Europe were displaced over their substratum by the given amount (crustal wandering of Europe to the north, contradicted by the distribution of the gravitational field strength); and (2) no internal axial shift of the earth occurred (improbable because of the systematic transgression cycle). This example—and many others could be cited in addition—shows how important a sharp definition of concepts is at this stage in the problem.

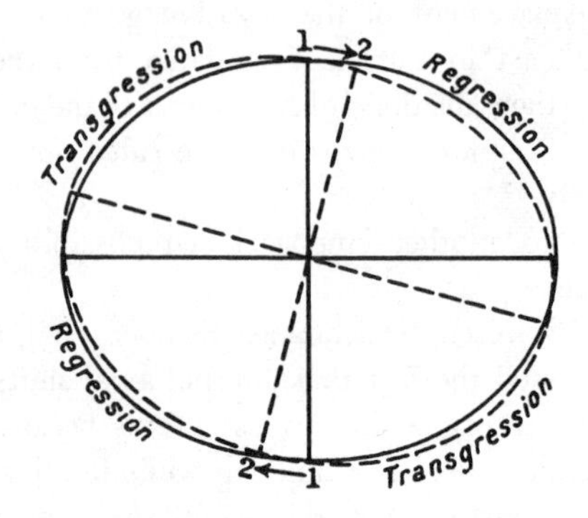


FIG. 41. Transgressions and regressions under conditions of polar wandering.

I have tried, though only provisionally, two methods of answering this question of transgressional cycles, using empirical data about them, and I can state in advance that both methods appear to confirm that internal axial shifts are associated with polar wandering.

The first test consists in a comparison of the transgressional change between the Devonian and the Permian with the simultaneous polar wandering that resulted. Strictly speaking, one should use the true polar shift; but the relative polar wandering used here, with Africa as the reference area, does not differ much from the absolute value. The largest uncertainty arises from the fact that the position and extension of the transgression seas at the various periods in question are only known very imprecisely.

Let us now mark on the reconstructed map of the world in the Carboniferous the coast lines of the transgression seas according to the usual palæogeographical representations, such as those of Kossmat or L. Waagen, for the two periods of the Lower Devonian and the Lower Carboniferous; this will give the areas which *became inundated* and those which *surfaced* in the interim, as shown in Figure 42. (These areas are not to be confused with the regions *lying* above or below water at the time.) In this period, however, the South Pole advanced from Antarctica towards South Africa,[38] so that South

[38] These figures are based on my earlier, provisional determination of the positions of the poles. The positions derived from more complete data in Köppen and Wegener, *Die Klimate der geologischen Vorzeit* [151], are somewhat different,

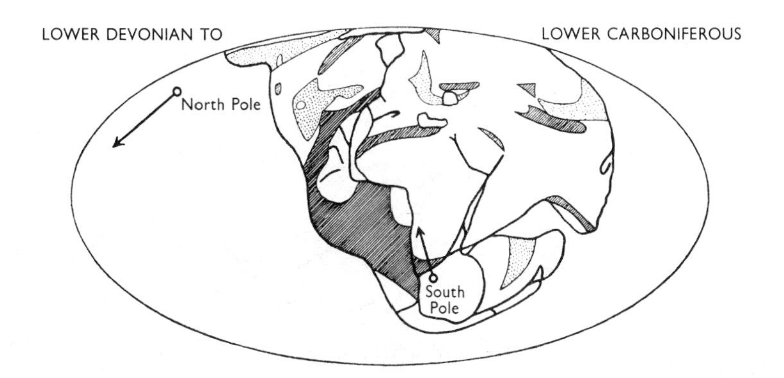


Fig. 42. Transgression (dotted), regression (hatched) and polar wandering between the Lower Devonian and the Lower Carboniferous.

America fell into the quadrant "in front" of the wandering pole. The North Pole, on the other hand, moved away from North America. We see therefore a confirmation of the rule: In front of the pole, regression; behind, transgression.

In the period which followed, from the Lower Carboniferous to the Upper Permian, the poles had a totally different direction of migration: The South Pole drifted from South Africa towards Australia, the North Pole again approached North America. Figure 43 shows the regions which surfaced and sank during this period, and again one can see confirmation of the rule, which appears all the more striking because the situation in both North and South America is exactly reversed.

These results thus appear to show that polar wandering from the Devonian to the Permian was actually connected with a displacement of the earth's axis on the inside.

I do not, of course, want to omit mention of the fact that the attempt to pursue this method of testing for the other periods of the earth's history has so far not led to unambiguous results. The next periods have, of course, such insignificant amounts of polar wandering that for this reason alone they appear to be little suited for such a test. Yet even for the Tertiary, with its large and rapid migration of the poles, I have so far obtained no clear results. Possibly one can no longer

but the difference is not enough to affect our conclusions. For this reason the figures were not corrected.

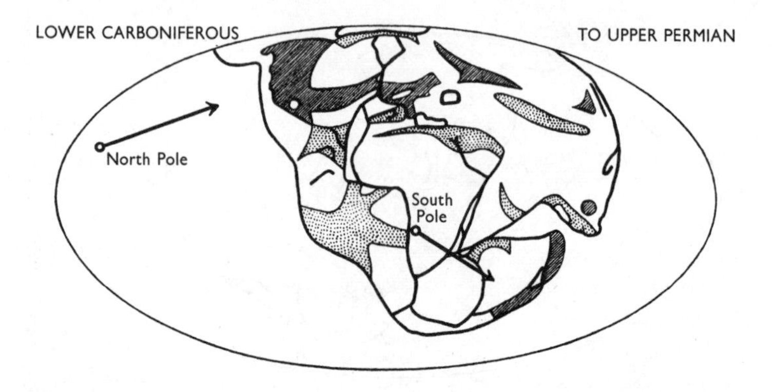


FIG. 43. Transgression (dotted), regression (hatched) and polar wandering between the Lower Carboniferous and the Upper Permian.

get by here with the relative polar wandering data that I used, and must base the investigation on counterbalanced polar wandering. But the greatest difficulty undoubtedly consists in the fact that the transgression seas for the individual subdivisions of the Tertiary, important here because of the rapidity of the variations, are mapped insufficiently or not at all. I presume, particularly in regard to what follows, that this is the reason why no clear picture emerges so far.

The second method of testing is this: Instead of the whole surface of the earth for a limited time span, one considers only a definite, thoroughly examined portion of the surface and sees how it behaved over the whole history of the planet (since the Carboniferous, in our case), comparing its changes of latitude with the cycle of transgressions. For if the rule "In front of the pole, regression; behind, transgression" is to hold, then each increase in latitude must be connected with a regression, and each decrease with a transgression. To test this, I have taken the best-known continent—Europe. For the variations in latitude we can use the figures (all are latitudes north) derived for Leipzig in Köppen and Wegener [151]:

| | |
|---|---|
| Carboniferous | 0° |
| Permian | 13° |
| Triassic | 20° |
| Jurassic | 19° |
| Cretaceous | 18° |

| | |
|---|---|
| Eocene | 15° |
| Miocene | 39° |
| Start of Quaternary | 53° |
| Now | 51° |

The latitude therefore increases from the Carboniferous up to the Triassic, then decreases up to the Eocene, then again increases up to the Quaternary. The highest latitude was attained by Leipzig in the Middle Quaternary.

On the other hand, geology tells us that, from the Carboniferous until the beginning of the Jurassic, regression generally prevailed over Europe; then large transgressions set in, forming the Jurassic sea and the Cretaceous sea and keeping a large part of Europe under water until as late as the beginning of the Eocene. From there on, a remarkable regression started again, as a result of which Europe as a whole became dry land. Even the final, slight fall in latitude value since the Quaternary has certain transgression phenomena that appear to correspond to it. At all events, the rule generally holds good, and this carries special weight here because Europe is the continent which has been best examined. This test, too, thus seems to show that polar wanderings are in fact connected with internal axial shifts of the earth.

In conclusion, we would like to touch briefly on the question whether the earth's axis undergoes and has undergone *astronomical* shifts also, i.e., whether variations have taken place relative to the system of the fixed stars.

It is known from astronomy that such variations occur at the present time. The precessional motion has been known for a very long time, by virtue of which the pole circles round the pole of the ecliptic every 26,000 years, without any change in the inclination of the earth's axis to the orbit, i.e., the angle of the ecliptic. The superimposed nutation is slight and is therefore not considered here. However, besides this, the calculations of perturbation show that the ecliptic angle, too, undergoes quasi-periodic oscillations of several degrees with a period of about 40,000 years; in spite of the small amplitude, these oscillations, in combination with corresponding variations in the size of the perihelion and the orbital eccentricity, were of decisive influence during the Quaternary in the genesis of the alternation of glacial and interglacial periods.

We may assume that these oscillations of the ecliptic angle have continued throughout the whole history of the earth and have thereby had an effect on the climate similar to that in the Quaternary period. For example, in the Permo-Carboniferous glaciation, traces have recently been found of repeatedly alternating advance and retreat of the ice, and further investigations will probably reveal more of them. It is very probable that this periodic oscillation of the ecliptic angle was a controlling influence in the origin of these advances and retreats, as were the corresponding oscillations of the Quaternary. The opinion has already been expressed that the apparently periodic variations in sedimentary deposition are connected with this variation in the ecliptic angle.

However, on the question whether even the mean value about which the ecliptic angle thus fluctuates has undergone considerable changes in the course of the earth's history, the astronomical perturbation calculation can give us no information, and that for two reasons. Firstly, the perturbation calculation involves the masses of all the planets of the solar system, and they are only known to a certain degree of accuracy; this makes the extrapolation of the calculation to geological periods (except the most recent, the Quaternary) quite illusory. Secondly, the earth is not a solid body, as is assumed in the calculation, but exhibits flow and is subject to continental drift, crustal wandering and probably also internal axial displacements; all these characteristics must have considerable influence on the result, but they cannot be taken into account in the calculation at the present time. We can therefore obtain no further information from this angle.

I would like, however, to draw attention to a peculiarity of the geological climates, of great interest in this connection. In the Permo-Carboniferous the South Pole region was in Gondwanaland and an inland ice formation existed there, at least equal to that of the present day. After this, we find throughout the following periods—Triassic, Jurassic, Carboniferous, up to the early Tertiary—no reliable traces of inland ice anywhere on the earth, although at least one of the poles was on land or near land most of the time, so that there can hardly have been any lack of opportunity to form ice sheets inland. At the same time, we find an astonishingly extensive advance of the plant and animal kingdoms towards the poles. Not until Tertiary times did new inland ice sheets form at the North Pole, reaching their greatest extent in the Quaternary. These variations in polar climate can be very well accounted for by the assumption that the mean value about which the 40,000-year period of the ecliptic angle fluctuates underwent considerable changes during the course of the earth's history, and that the process involved a small ecliptic inclination in times when inland ice was present, and a large one at times when there was no ice and the organisms made large advances.

It is, of course, not difficult to understand the effects of such variations in ecliptic angle on the earth's climatic system. One needs only to realise that the annual temperature variation basically derives from the ecliptic angle. If this were zero, the earth's axis would be normal to the solar orbit, and with the small orbital eccentricity that obtains, the annual variation would as good as disappear, and everywhere on the earth, throughout the year, the temperature would remain constant on time, which is the case today only in the tropics. The (very low) mean temperature of the polar regions would prevail there all through the year; the winter would certainly be warmer than now, but the temperature would remain permanently below the freezing point. Summer would be indistinguishable from winter. Plant life would then be out of the question, since there would be no period of growth during the year. The plant kingdom would therefore be forced a long way back from the poles, and the terrestrial animals would have to follow. All precipitation would be in the form of snow throughout the year, and could never melt because the lack of a summer period means lack of a period for melting. Snow would then accumulate and all the land would be covered with inland ice.

If, on the other hand, the ecliptic angle were appreciably larger than today's, the annual variation in temperature at the poles would increase enormously. The polar summer would be much warmer, so that plants, and with them, terrestial animals, would be able to colonise the whole region inclusive of the pole; even tall trees would be able to grow there if the mean temperature of the warmest month exceeded 10°C because, as shown in Siberia, many forms are able to survive severe winters. Summer precipitation would fall as rain, while the winter precipitation, falling as snow, would easily melt in the summer heat, so that no inland ice could form, even with low mean annual temperatures, as is the case in Siberia. Further, the mean annual temperature in polar regions would rise, even if only slightly, because the more intense solar radiation of the summer period could not be fully offset by greater radiative heat loss in winter; for if the sun is only just below the horizon, as far as the radiation balance is concerned, that is the same as if it were far below. The evidence for climatic conditions as provided by the plant and land-animal kingdoms of such periods will of necessity give the impression of a moderation in climatic differential between pole and equator.

Admittedly, the above-mentioned palæoclimatic evidence for such variations in polar climate during the earth's history still needs much more extensive investigation. One should also note that still other causes can be found for these variations. At the present time, however, it does seem likely to me that they are real oscillations, and that they can best be explained by changes in the ecliptic angle. They would therefore be indications that, besides the astronomical axial variations of the earth known so far, still others have occurred which elude astronomical computation.

CHAPTER 9

# *The Displacement Forces*

THE determination and proof of relative continental displacements, as shown by the previous chapters, have proceeded purely empirically, that is, by means of the totality of geodetic, geophysical, geological, biological and palæoclimatic data, but without making any assumptions about the origin of these processes. This is the inductive method, one which the natural sciences are forced to employ in the vast majority of cases. The formulation of the laws of falling bodies and of the planetary orbits was first determined purely inductively, by observation; only then did Newton appear and show how to derive these laws deductively from the one formula of universal gravitation. This is the normal scientific procedure, repeated time and again.

The Newton of drift theory has not yet appeared. His absence need cause no anxiety; the theory is still young and still often treated with suspicion. In the long run, one cannot blame a theoretician for hesitating to spend time and trouble on explaining a law about whose validity no unanimity prevails. It is probable, at any rate, that the complete solution of the problem of the driving forces will still be a long time coming, for it means the unravelling of a whole tangle of interdependent phenomena, where it is often hard to distinguish what is cause and what is effect. It is clear from the start in this question of the forces that the whole complex of continental drift, crustal wandering, polar wandering, internal and astronomical shifts forms a unified problem.

So far, only a single aspect of the question has been solved, and conjectures have been advanced about a few others.

On the question of the forces producing drift, of special interest in the first instance are those movements which we have designated previously as crustal wanderings, i.e., displacements of the continental blocks relative to their substratum. These are of interest because, at least in the majority of cases, they should be regarded as direct effects of displacement forces which are applied to the continental blocks, but which act on the underlying material to a lesser extent or not at all.

We have already referred to a large number of details which constitute evidence of these two forms of displacement. The westward drift of the continental blocks is immediately evident when one looks at a present-day map of the world. The "flight from the poles," so far as earlier movements are concerned, is largely concealed by the present, altered position of the poles, and appears in its proper light only after reconstruction of the positions of the poles at the time in question. However, the drift from the poles is exhibited in a general way by the splitting up of continental blocks in the polar regions and the compressive thrust which acts on them towards the equator. For instance, the Permo-Carboniferous advance of the South Pole towards Africa was accompanied by the Carboniferous folding process along the equator as it was then, and was followed by a disruption and dispersion of Gondwanaland. In just the same way, the Tertiary advance of the North Pole, formerly lying in the Pacific, towards the land mass of the present-day north polar region was accompanied by the Tertiary folding process along the equator of that period (the Alps to the Himalaya); this was followed, and is still being followed, by increasing disruption and dispersion of the northern continents.

The only displacing force of which we have precise knowledge today is that producing a "flight from the poles." This force acts to drive the continents over their substratum towards the equator. As far back as 1913, Eötvös stated its existence in a comment [199] which, to be sure, passed unnoticed at that time. In a discussion of the subject, he drew attention to the fact "that the direction of the vertical in the plane of the meridian is curved, so that the concave side faces the pole, and that the centre of gravity of the floating body (the continental block) lies higher than that of the mass of liquid displaced. Thus, the floating body experiences two forces acting in different directions, the resultant being directed towards the equator and away from the pole. The continents would therefore tend to move towards the equator, and this movement would give rise to a secular variation of latitude, as surmised in the case of the Pulkovo observatory."

Unaware of this short, inconspicuous reference, W. Köppen [200] recognised the nature of the pole-flight force and its significance for the question of continental drift; he gave a description of the force, though without quantitative analysis:

"The flattening of the surfaces at any given level therefore decreases with depth; the surfaces are not mutually parallel but are slightly tilted with respect to each other, except at the equator and poles, where they are normal to the radius of the earth. The figure [Fig.

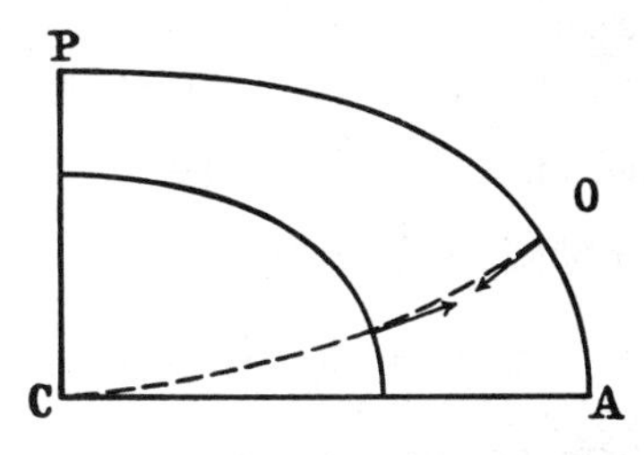


FIG. 44. Two surface levels and the curved vertical.

44] shows this at a meridional cross section between pole (P) and equator (A). The dashed line concave towards the pole is the line of force of the gravitational field or the vertical at position O. C is the centre-point of the earth.

"The point of application of the buoyancy force acting on a floating body lies at the centre of gravity of the displaced medium, but the point at which the body's own weight acts is at its own centre of gravity, the direction of both forces being normal to the surface at the point in question; the force vectors are therefore not opposed, but have a resultant of small magnitude, which, if the buoyancy centre lies below the centre of gravity, points towards the equator. Since the centre of gravity of the block is far below its surface, the two forces are not directed along the normal to the plane of its surface, but slightly inclined to it, the buoyancy force more so than the gravitational pull on the block. These laws must apply for any floating body whose centre of gravity is higher than its centre of buoyancy, whereas the forces must have a resultant directed towards the pole if the body's centre of gravity is below the buoyancy centre. Archimedes' principle is strictly correct for a rotating earth only if both centres coincide."

The first calculation of the force directed away from the poles was carried out by P. S. Epstein [201]. For the force $K_\phi$ at latitude $\phi$ he derived the function

$$K_\phi = -\frac{3}{2}(md\omega^2 \sin 2\phi),$$

where $m$ is the mass of the continental block, $d$ half the difference in height between sea floor and block surface (or the difference in height between the centre of gravity of the block and of the displaced sima) and $\omega$ the earth's angular velocity.

Epstein used this equation to calculate the coefficient of viscosity $\mu$ of the simasphere from the rate of drift $v$ of the continental blocks (according to the equation $K = \mu \frac{v}{M}$, where $M$ is the thickness of the viscous layer). He obtained the expression

$$\mu = \rho \frac{sdM\omega^2}{v},$$

where $\rho$ is the specific gravity of the block and $s$ its thickness. He then used the following numerical values to find $\mu$:

$$\rho = 2.9; \quad s = 50 \text{ km}; \quad d = 2.5 \text{ km}; \quad M = 1600 \text{ km};$$

$$\omega = \frac{2\pi}{86{,}164}; \quad v = 33 \text{ m/yr};$$

he found the viscosity coefficient of the sima to be

$$\mu = 2.9 \times 10^{16} \text{ g}\cdot\text{cm}^{-1}\cdot\text{s}^{-1},$$

which is three times that of steel at room temperature. If one takes $v$ as equal to 1 m/yr, which is probably nearer the true value, $\mu$ comes out 33 times larger, i.e., about $10^{18}$. Epstein concludes from this: "We can sum up our results by saying that the centrifugal force due to the earth's rotation can produce and must produce a flight from the poles of the amount proposed by Wegener." However, Epstein believes that the equatorial fold ranges cannot be ascribed to this force, since it corresponds to a surface drop of only 10 to 20 m between pole and equator, whereas the piling up of the mountains to several km altitude and the corresponding subsidence of sialic masses to great depths represent a large amount of work against the force of gravity, for which the pole-flight force is insufficient. The latter force could only form mountains of 10 to 20 m in height.

Almost simultaneously with Epstein, W. D. Lambert [202] derived the value of the pole-flight force mathematically, obtaining essentially the same result. He finds the force at 45° latitude, to be $\frac{1}{3 \cdot 10^6}$ of the gravitational force. Since the force reaches its highest value at this latitude, it must act to rotate any oblong continent which lies at an oblique angle; between the equator and 45° latitude, such a continent will be forced to align its long axis east–west, whereas between 45° and the pole the alignment will be along a meridian. Lambert says: "All this is quite speculative of course; it is based on the hypothesis of floating continental masses and on the assumption of a sustaining magma that would of course be a viscous liquid, but viscous in the sense of the classical theory of viscosity. According to the classical theory a liquid, no matter how viscous, will give way before a force, no matter how small, provided sufficient time be allowed for the latter to act in. The peculiarities of the field of force of gravity will give us minute forces, as we have seen, and the geologists will doubtless allow us æons of time for the action of the forces, but the viscosity of the liquid may be of a different nature from that postulated by the classical theory, so that the force acting might have to exceed a certain limiting amount before the liquid would give way before it, no matter how long the small force in question might act. The question of viscosity is a troublesome one, for the classical theory does not adequately explain observed facts and our present knowledge does not allow us to be very dogmatic. The equatorward force is present, but whether it has had in geologic history an appreciable influence on the position and configuration of our continents is a question for geologists to determine."

Schweydar also calculated the pole-flight force [40]. For a latitude of 45°, he obtained a velocity of about $\frac{1}{2000}$ cm/s, that is, the force amounts to about $\frac{1}{2 \cdot 10^6}$ of the weight of the block. He states: "Whether this force is enough to cause a displacement is not easy to say. In any case, it would not explain a westward migration since the speed is too low to give rise to a noticeable westerly deflection through the rotation of the earth."

Schweydar takes exception to Epstein's calculation on the grounds that the assumed displacement velocity (33 km/yr) is too high, and the sima viscosity derived therefrom much too low. But if the velocity is assumed to be smaller, the required higher viscosity is obtained. He says: "If one assumes a viscosity coefficient of the order of $10^{19}$ (instead of Epstein's $10^{16}$) and presupposes that the formula Epstein used is applicable here, one obtains for the block velocity, at latitude 45°, about 20 cm/yr. *Nevertheless, it must be stated as a possibility that the continents undergo a displacement towards the equator under the action of the force causing flight from the poles.*"

Finally, Wavre [204] and Berner [203] have carried out a new computation of the pole-flight force, probably the most accurate one. They obtain as a maximum value, for 45° latitude, $\frac{1}{800,000}$ of the weight of the block. They state: "The ratio of displacing force to weight of the continent is therefore extraordinarily small; it could not produce mountain ranges and at the present time is not producing them at the equator.

"However, matters are otherwise if a dynamic effect is added to this static one.

"The resistance of the sima does not prevent the continents from moving; and in the case where two continents meet at the equator or in other latitudes, the kinetic energy lost by either one must be recovered in one form or another."

It would appear that Kreichgauer was the first to discover the force underlying the flight from the poles. In the second edition of his book *Die Äquatorfrage in der Geologie* [5] he inserted on page 41 a consideration he had already published elsewhere in 1900, which yields the pole-flight force. In the first edition this is missing.

I would like to mention further that M. Möller [205] also, in 1922, published a derivation of the pole-flight force which he developed in 1920.

Probably this literature survey could be amplified; I have only cited those items which I have come upon by chance.

If we therefore assume, with Wavre and Berner, that the pole-flight force is about $\frac{1}{800,000}$ of the weight of the continental block, we should nevertheless observe that this is about 15 times the horizontal tidal force; and while the latter continually changes direction, the pole-flight force acts in the same direction and with the same strength for millennium after millennium. It is this which enables it to overcome the earth's viscosity, similar to that of steel, in the course of geological time.

Recently, Lely has made an interesting attempt to demonstrate the force directed away from the poles [206]. I have repeated this with

J. Letzmann, and we found that it serves as an excellent lecture demonstration. A cylindrical water vessel is placed on a stool turntable and very accurately centred. When the water is brought into smooth rotation with the vessel, its surface takes on paraboloidal curvature (Fig. 45*a*). A float is now placed on the water surface, consisting of a flat cork with a nail stuck in the centre (Fig. 45*b*). The nail should

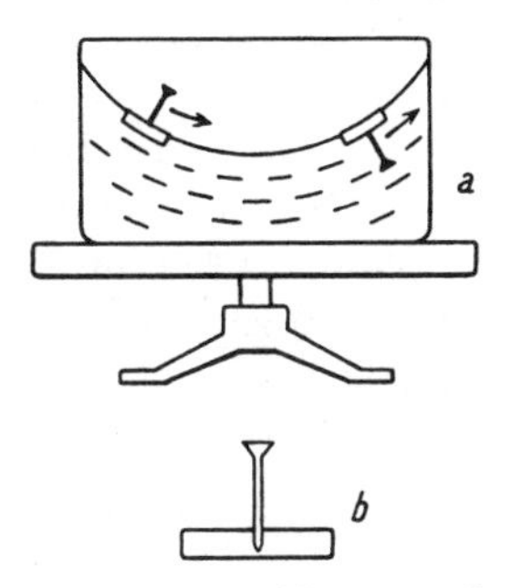


FIG. 45. Lely's experiment to illustrate the pole-flight force.

be as long as possible, but the cork should still be able to float with the nail upright and not turn over. The float is now set on the rotating surface of the water, with the nail first upwards, then downwards. When the nail is pointing upwards, the float is observed to drift quickly to the centre; with the nail pointing downwards, however, it drifts to the side. If the float is placed on the water several times in succession in alternate positions, so that the direction of movement changes each time, the experiment has a very convincing effect.

The fundamental explanation of this experiment is very simple if one remembers that the centre of gravity of the float does not coincide with that of the water it displaces, but when the nail is upright lies above, and when the nail points downwards lies below, that of the displaced water. As shown by the curvature of its surface, a radial pressure gradient prevails in the water, exactly compensated by centrifugal force. Were the centre of gravity of the float to coincide exactly with that of the displaced water, no displacing force would arise, because the float would be subject to the same condition; the pressure differential across the outer and inner lateral surfaces would be precisely counterbalanced by centrifugal force. However, if, when the nail is upright, the float's centre of gravity is displaced upwards and in a direction normal to the water surface, it will be simultaneously moved nearer to the axis, the centrifugal force will decrease and the excess pressure gradient will drive the float to the centre. Conversely, when the nail points downwards, the float must drift to the edge, because its centre of gravity is farther from the axis than that of the water it has displaced, so that now the centrifugal force outweighs the pressure gradient.

At first sight, this experiment appears to demonstrate just the opposite of the pole-flight force, because the continents, with their centres of gravity above those of the material they displace, correspond to the float with upright nail. One can easily see, however, that this reversal is simply a consequence of the fact that the curvature of the "fluid" is in the opposite sense. The centre of gravity of the continents lies farther from the axis than that of the sima displaced, just because of the convex curvature of the earth's surface, while in the experiment its distance from the axis is relatively less.

It should be clear from the foregoing that the pole-flight force is enough to displace the continental blocks through the sima, but not enough to produce the great fold-mountain ranges that arose in association with the flight of the continents from the poles. Of course, Berner has rightly pointed out that this only proves true if one considers the static pressure experienced by a stationary continental block in a horizontal direction, by virtue of the pole-flight force. It is quite a different thing if we assume, for example, that under the action of the pole-flight force, which must overcome the viscous drag of the magma, a large continental block drifts towards the equator at a uniform rate, and only then, in the course of this movement, encounters an obstacle which applies a braking force. Moreover, the block must come to rest; this destroys its kinetic energy. Admittedly, one should not overestimate this effect. The kinetic energy is half the mass times the square of the speed. It is true that the moving mass is very large, but the speed whose square is involved is very low, so that as a rule it is not possible to explain the formation of mountain ranges in this way, either. We must therefore still assume that the ordinary pole-flight force is an insufficient explanation for orogenesis.

For some strange reason, some geologists appear to regard this situation as an objection to drift theory; this is quite illogical. The existence of the fold ranges cannot be doubted. If they demand a force larger than the pole-flight force, their existence is evidence that, in the course of the earth's history, at least at times, displacement forces have arisen which were much larger than the pole-flight force. But if this force is enough to cause continental drift, the unknown orogenic forces must have been all the more able to do so!

We can be much briefer in discussing the forces which could account for the westward continental drift. Various authors, such as E. H. L. Schwarz, Wettstein, *et al.*, have put forward a claim for *the friction of tidal waves* as the driving force producing a rotation of the whole earth's crust over the core in a westerly direction. These waves are produced in the solid body of the earth by the gravitational attraction of sun and moon. It is often assumed that the moon once rotated more rapidly, but that the tidal friction produced by the earth had a braking effect. However, it is easy to see that this retardation of a heavenly body by tidal friction must chiefly affect its outermost layers, leading to a slow sliding motion of the whole crust or of the individual continental blocks. It is just a question of whether such tides really exist. According to Schweydar, the tidal deformation of the solid earth, detectable by means of the horizontal pendulum, is of a different nature, that is, elastic. This cannot therefore be directedly adduced in explanation. However, W. Lambert [221] is of this opinion: "Nevertheless, we cannot believe that the free oscillation is wholly unaffected by frictional resistance, although this cannot be reliably detected by the observations." It is, in fact, quite beyond question that we cannot regard the earth as completely elastic with respect to the tidal forces. There must, therefore, be tidal flow also, as well as the elastic and measurable tides; the tidal flow is, to be sure, below the limits of measurement because the period of this flow is too short relative to the viscosity of the magma, but in the course of geological time the tidal friction effect accumulates and can ultimately produce considerable displacements of the earth's crust. My view at any rate, is that one cannot yet regard it as settled that the elastic nature of the measurable, daily tides in the solid earth is a demonstrable fact.

Schweydar arrived at a force that could lead to a westward drift of the continents by another approach, which, however, also depends ultimately on solar and lunar attraction [40]; this approach derives from the theory of the *precession* of the earth's axis. Schweydar states: "The theory of the precession of the rotational axis of the earth under the influence of the attractive forces of sun and moon presupposes that the individual portions of the earth cannot undergo large mutual displacements. The calculation of the shift of the earth's axis in space is more difficult if the continents are assumed to have drifted. In this case, a distinction must be made between the axis of rotation of the continents and that of the earth as a whole. I have calculated that the precession of the rotational axis of a continent lying between latitudes −30° and +40° and meridians 0° and 40° W is about 220 times larger than that of the axis of the whole earth. The continent will tend to rotate about an axis which differs from the general axis of rotation. Hence forces arise which act not only in a meridional direction, but also in a westerly one, and which seek to displace the continent; the meridional force reverses its direction in the course of the day and does not enter the problem. These forces are much larger than that producing flight from the poles. The force is strongest at the equator and zero at ±36° of latitude. I hope later to give a more precise description of the problem. As a result of the theory, a westerly displacement of the continents would be possible." Although this, too, is only a provisional statement (the promised final version has unfortunately still not appeared), it nevertheless seems very likely that the most clearly recognisable general movement of the continents, their westerly drift, could be explained by the attractive forces of the sun and moon acting on the viscous earth.

However, Schweydar thinks that the deviations of the earth's *shape* from the ellipsoid of revolution, deduced from measurements of gravitational field strength, can give rise to flow in the sima and therewith to drift of the continents: "One can also conjecture that there have been flow movements in the sima, at least in earlier epochs. In his latest work on the distribution of gravitational field strength over the earth's surface, Helmert concluded that the earth is a triaxial ellipsoid; the equator forms an ellipse. The difference between the axes of this ellipse is only 230 m; the major axis cuts the earth's surface at 17° W (the Atlantic), the minor at 73° E (the Indian Ocean). According to the theories of Laplace and Clairaut, which have not been superseded in geodesy, the earth is regarded as a fluid, i.e., the pressure in the solid earth (except for the crust) is assumed to be hydrostatic. From this viewpoint, Helmert's result is unintelligible. A hydrostatic earth, considering its oblateness and angular velocity, cannot be a triaxial ellipsoid. One could then assume that the divergence from an ellipsoid of revolution derived from the continents. But that is not the case. I carried out the calculation on the assumption that the continents are floating and have the thickness quoted above [200 km; density difference between sial and sima = 0.034 (water = 1)]; I found that the distribution of the continents and oceans gives rise to a difference beween the mathematical shape of the earth and an ellipsoid of revolution much smaller than that found by Helmert. Besides this, the axes of the equatorial ellipse are very differently situated from the positions given by Helmert; the major axis comes out in the Indian Ocean. It must be the case, therefore, that large portions of the earth do not behave hydrostatically.

"According to my calculation, Helmert's result can be explained if a 200-km thickness of sima under the Atlantic is denser by 0.01 than it is under the Indian Ocean. Such a condition cannot be sustained indefinitely, and the sima will tend to flow to restore the equilibrium state of the ellipsoid of revolution. Because of the small density difference, hardly any flow would be possible, but the ellipticity of the equator and the difference in density in the sima, and therefore the flow, could have been more significant in earlier epochs."

It will be clear at once that the forces deduced from Helmert's result can serve to explain the opening up of the Atlantic Ocean, since just here the earth seems arched and the masses are probably constrained to flow apart on either side.[39]

Another consideration should be brought up at this point, which one may perhaps consider as an extension of the train of thought discussed so far. Such an arching of the earth's surface above its equilibrium position naturally need not be confined to the equator, but can occur anywhere on the earth. It was shown previously when discussing the transgressions and their connection with polar wandering (Chapter 8), that in front of a shifting pole we must expect too high a position of the earth's surface, and behind it one that is too low, and that the geological facts appear to confirm the presence of these deviations. Here too, we are dealing with amounts similar to the one Helmert found for the excess of the major over the minor equatorial axis, or perhaps double that amount. When polar migration is more rapid, the earth's surface appears in any case to be a few hundred meters above the equilibrium level in front of the pole, and a few hundred metres below it behind the pole. The largest gradient (of the order of 1 km per quadrant of the earth) would exist in the meridian of the polar displacement at its point of intersection with the equator, and there would be one almost as large at each of the poles. Forces are released thereby which drag the land masses from the excessively high

[39] It should be pointed out, however, that doubt has been expressed recently that the earth really is a triaxial ellipsoid. Heiskanen found that this result is simulated merely by unfavourable combinations of gravity measurements [219].

to the excessively low regions, and these forces are many times the normal pole-flight force, which, in the case of the continental blocks, corresponds only to a gradient of 10–20 m per quadrant. These forces, unlike the pole-flight force, do not act *only* on the continental blocks, but also on the underlying sima, which is more fluid and may perhaps restore equilibrium under the solid crust. But so long as the gradient exists—and the transgressions and regressions appear to testify to its existence—this force must act on the continental blocks, and it must therefore also be able to produce displacements of them and folds in them, although these movements may be less than the corresponding movements of the more fluid material beneath them. I would like to believe that we have in this deformation of the earth's shape by polar wandering a completely adequate source of power to supply the energy required for folding.

This intepretation is rendered specially probable by virtue of the above-mentioned fact that the two largest fold systems involved here, namely the equatorial folds of the Carboniferous and the Tertiary, originated in just those eras for which other reasons lead us to assume that polar wanderings were particularly rapid and extensive.

Recently, several authors, such as Schwinner [69] and especially Kirsch [70], have made use of the concept of convection currents in the sima. In conjunction with Joly's idea that the sima under the continental blocks is heated by the large radium content, and that in oceanic regions it cools, Kirsch assumes a circulation of sima beneath the crust: It rises below the continents up to their lower boundary, then flows along under them to the ocean regions, where it flows downwards, returning to the continents after reaching greater depths. Because of the resulting friction, he says, the sima tends to disrupt the continental cover and to force the fragments apart. We mentioned earlier that the relatively great fluidity of the sima, as assumed here, has been regarded as unlikely by the majority of authors to date. In considering the earth's surface, however, there is no mistaking that the split-up of Gondwanaland and also that of the former single continental block composed of what is now North America, Europe and Asia, can be conceived as the effect of such sima circulation. This idea also apparently offers a reasonable explanation of the opening up of the Atlantic Ocean. It cannot therefore be utterly rejected on the grounds that the superficial phenomena of the earth would gainsay it. If the theoretical basis of the ideas should prove adequate to support them, they could in any case be considered as contributory factors in the formation of the surface of the earth; it is still not possible at present to survey the theoretical background.

Our discussion will have shown the reader that the problem of the forces which have produced and are producing continental drift (except the pole-flight force, already thoroughly investigated) is still in its infancy.

We may, however, assume one thing as certain: *The forces which displace continents are the same as those which produce great fold-mountain ranges.* Continental drift, faults and compressions, earthquakes, volcanicity, transgression cycles and polar wandering are undoubtedly connected causally on a grand scale. Their common intensification in certain periods of the earth's history shows this to be true. However, what is cause and what effect, only the future will unveil.

CHAPTER 10

# *Supplementary Observations on the Sialsphere*

THE MAIN evidence in support of drift theory has been discussed in the previous chapters, and this being so, we now prefer to assume its validity, and in this and the subsequent chapter we shall present, by way of a supplement, a number of phenomena and problems so closely tied up with our theory that some discussion of them seems desirable. I want to stress that these accounts are aimed more at raising questions and providing a stimulus to discussion than at proposing definitive solutions.

Let us first consider the sialsphere, which today provides only a fragmentary cover of the earth in the form of the continental blocks.

Figure 46 gives a map of the world's continental blocks. Since the continental shelves belong with them, the outlines diverge considerably in many places from the well-known coastlines. For our considerations it is important to free oneself from the usual picture of the world map and to gain a certain degree of familiarity with the outlines of the entire continental blocks. As a rule, the 200-m depth contour best represents the edge of these tables, but there are some parts, certainly still belonging to the blocks, which reach a depth of 500 m.

We said earlier that the material of these continental blocks is mainly granite. It is well known, however, that the surface of the blocks is to a great extent not composed of granite, but of sedimentary deposits, and we must therefore become clear in our minds what rôle these play in the structure of the blocks. One may take 10 km (approx.) as the greatest thickness of the sediments, a value computed by American geologists for the Palæozoic sediments of the Appalachians; the other thickness limit is zero, since in many places the primitive rock is bare of any sediment covering. Clarke estimates the mean thickness on the continental blocks as 2400 m. Since, however, the present-day overall thickness of the blocks is estimated at about 60 km, that of the granite layer at about 30 km, it is clear that this sedimentary cover represents merely a superficial layer produced by weathering; moreover, should it be completely removed, the blocks would rise virtually to their former elevations to restore isostasy, and the relief of the earth's surface would be little altered.

The map (Fig. 46) should not be read as implying that the block margins, indicated by thick lines, mark the boundary between sial and sima. As will be shown in the next chapter, the ocean floors are probably still covered with sial remains in many places. By the term "continental blocks" one should understand the still intact, essentially unbroken sial covers in contrast to those oceanic sial masses which in their form represent broken fragments of the blocks, the results of surface break-up and, in deeper layers, of the pulling apart or the drifting apart of the material. One must therefore distinguish between the general concept of the sial cover and the more specialised one of sial blocks. Only the latter are represented in our map.

The most radical changes undergone by these sial blocks in the course of geological time are undoubtedly the alternating transgressions (inundations) and regressions (drainage), whose behaviour is connected with the accidental circumstance that the amount of water in the oceans of the world is just a little more than the available ocean basins can contain, with the result that the lower-lying portions of the continental blocks still lie under water. If the sea levels of the world were 500 m lower, these phenomena, which play so outstanding a part in geology, would be confined to narrow marginal strips. The present-day transgressions can be seen immediately from our map. Small variations in level of the block surfaces therefore result in large displacements of these inundated areas.

Generally speaking, we are dealing here with variations in water level which do not exceed a few hundred metres in height. Past transgression seas were as shallow as those of today. The question arises: How do these well-established level changes conform to the principle of isostasy or to the hydrostatic equilibrium of the earth's crust? The probable answer is: If a continental block is forced below its hydrostatic equilibrium level by any influence, a mass deficit naturally arises which calls into play forces which tend to restore the equilibrium position. So long as the changes in level remain within

FIG. 46. Map of the continental blocks (Mercator's projection).

given limits, the gravitational anomaly will remain within the limits which are actually observed at various points on the earth as small regional deviations from isostasy. Since the material is very viscous, it is clear that a certain threshold value of change in level needs to be exceeded before the forces become so strong as to produce appreciable isostatic equilibration movements. It is therefore possible that this amount of a few hundred metres roughly represents this threshold value, which naturally cannot be considered an absolute constant.

The explanation of the cause of transgression cycles in the history of the earth will represent one of the most important, but also one of the most difficult tasks of future geological and geophysical research. At the present time, the question can still not be regarded as solved, although notable beginnings have been made towards at least a partial solution. The chief difficulty at the moment is that the geological surveys—in spite of the many palæographical world-maps—are not nearly reliable or complete enough to permit empirical tracing of these transgression cycles with regard to location and date, so that the material at hand is mostly insufficient to test the hypotheses put forward in explanation. However, one can say at this time that the totality of transgression cycles certainly cannot be attributed to one single cause, because various causes could be mentioned which must be considered as at least contributing factors, so that certainly the problem is inherently complex. Naturally, that does not rule out the possibility that at some future time one cause will be recognised as overriding.

So far, to the best of my knowledge, the following causes can be adduced:

1. An appreciable change in the amount of water contained by the oceans of the world, produced by formation and melting of large inland ice masses, must naturally lead to a change in the extent of transgression. These transgressional alternations must have the characteristic of proceeding in the same kind of way over the whole earth and without disturbance of the isostatic state. It can easily be calculated that formation of an ice cap as large as those of the Quaternary or Permo-Carboniferous will produce a drop in sea level of about 50–100 m.[40]

2. Elevations and depressions of the sial surface can also result, without isostatic disturbance, from horizontal compression (orogenesis) or from horizontal extension (fault formation at the surface,

[40] cf. Born in [45], p. 141.

withdrawal of the deeper layers) of the sial cover. The thickness of sial cover is increased in the first case, decreased in the second. For example, the Alps were raised above sea level by folding, while the Aegean Sea region was sunk, with the formation of many faults, till only the islands remained (cf. Fig. 47, a diagrammatic illustration).

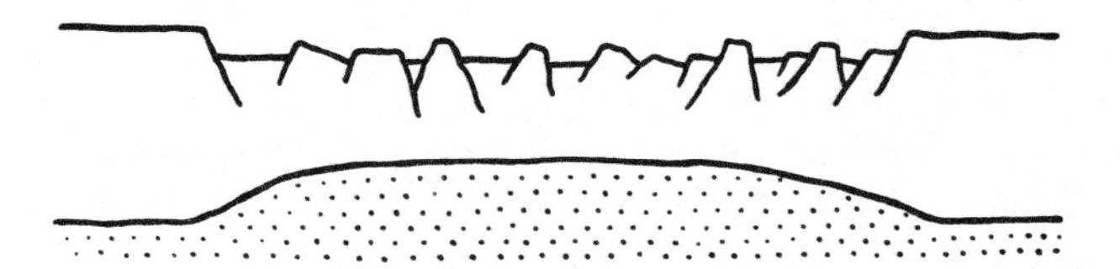

FIG. 47. Extensive break-up of sial through elongation of substratum (diagrammatic).

Although one may encounter very considerable local gravitational anomalies, these processes do not basically involve disturbances of isostasy, or at least not disturbances which correspond to the amount of elevation of subsidence in question. Further, they are associated with large variations in the horizontal dimensions of the regions they affect, and overall they exhibit a local rather than regional character.

3. Other causes can also be astronomical variations in the earth's motion, especially such as affect the equilibrium oblateness of the planet. The ocean will follow the latter type of variation *without* any time lag, but the very viscous earth will accommodate *with* a time lag; if the flattening increases, transgressions must arise at the equator and regressions at the poles; if it decreases, the reverse will hold. Amongst other causes, such oblateness changes may derive from variations in the angular velocity of the earth, as recently determined by observations (although their interpretation as such is still uncertain!), and also from variations in the ecliptic angle; this is because if the angle is large, tidal forces must produce a definite, though slight, elongation of the earth's shape in the direction of its axis. The converse occurs if the angle decreases—there is an increase in the equatorial radius. Thus, if the angle of tilt increases, transgressions are to be expected at the poles, but regressions if it decreases, the converse holding at the equator.

4. In so far as the geologically determined polar migrations signify that a shift of the earth's axis relative to the whole earth has occurred,

they must be a very fertile source of transgressional changes, as discussed in the previous chapter. As was shown, the phenomena in fact imply the reality of this effect, whereby increasing regressions appear to prevail in front of the migratory pole, and transgressions behind it. I think it not unlikely that this will prove to be the prime cause of transgressions; however, what has been said indicates that there are still other causes to be considered, the number of which may even be augmented.

The phenomena, discussed in point 2—elongation to fracture-point and compression folding—form the second class of main events to which the continental blocks are subject, besides the transgressional changes. They have long been objects of study in tectonics. We only want here to cite a few points of interest in this connection. It has been known for a long time that fold ranges are formed under considerable horizontal compression, although even today this is disputed by a few authors who would like to account for fold mountains by a fundamentally different process; they represent such an isolated group in this respect that we need not discuss the question further here. What is important is that, in the case of both ancient as well as recent fold ranges, we find no gravitational anomalies of such magnitude as would have to be present if these chains of mountains had simply been laid down on the earth's crust. It is true that one often finds in such ranges quite measurable deviations from complete isostasy, and discussion of these is of great interest in other respects, but they are so small that in a first approximation we can say: Folding of the mountain chains is carried out essentially with conservation of isostasy. The meaning of this is explained in the diagram of Figure 48. When a continental block floating on sima is compressed, the ratio of what is above to what is beneath the surface of the sima must always remain

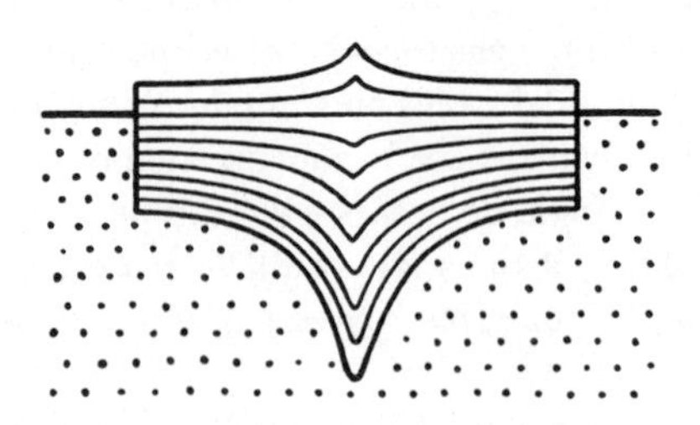

Fig. 48. Compression with conservation of isostasy.

the same. Depending on whether we assume the thickness of the sial cover, which projects 5 km from the sima, as 30 or 60 km, we obtain the ratio 1 : 6 or 1 : 12. Therefore, the downward-directed portion of the compression must be respectively 6 or 12 times the upward-directed portion. Thus, what we see of the mountain range is but a small fraction of the whole compressed mass. In the case of an ideal compression, we see only those layers which were already above sea level before the compression occurred. Whatever lay below this level, remains below after or during compression, if one ignores minor perturbations. Therefore, if the superstructure of the block consists of a 5-km thickness of sedimentary "skin," then the whole range would have originally consisted of sediment alone. Only when this is removed by erosion does a central chain of primitive rock rise until isostatic equilibrium is restored, until finally, once the entire sedimentary cover has been swept away, a wide primary range of almost the same mean elevation has risen up. The Himalaya and neighbouring ranges would rank as an example of the first stage. The erosion of these sedimentary folds is so extensive that the glaciers are almost buried under the moraines; a case in point is the Baltoro glacier, the largest in the Karakorum range, which, while a mere $1\frac{1}{2}$ to 4 km wide (65 km long), has no less than 15 medial moraines. The second stage, where the central chain already consists of primitive rock, but is still flanked on both sides by sedimentary zones, is represented by the Alps. Since erosion in the primitive rock is much slighter, the Alpine glaciers have few moraines, a primary cause of their beauty. The Norwegian ranges represent the third, final stage. The sedimentary cover is here for the most part totally removed, and the rise of the primary range is complete. Thus the range is also cleared of sediment under conservation of isostasy.

Very often one can recognise that the parallel folds of a range are staggered (in echelon). If one pursues such a fold, one finds that it emerges sooner or later from the edge of the range and finally disappears; the next chain inwards then forms the margin until it, too, disappears some distance farther along, and so on. This occurs when the two blocks do not move exactly towards each other, but undergo a shearing motion, though with a component along their mutual perpendicular. In general, the effects of the various types of movement of blocks relative to each other can be elucidated by means of Figure 49. Here, the left-hand block is stationary, the right-hand moves. If its movement is normal to the boundary between blocks, no echelon folds

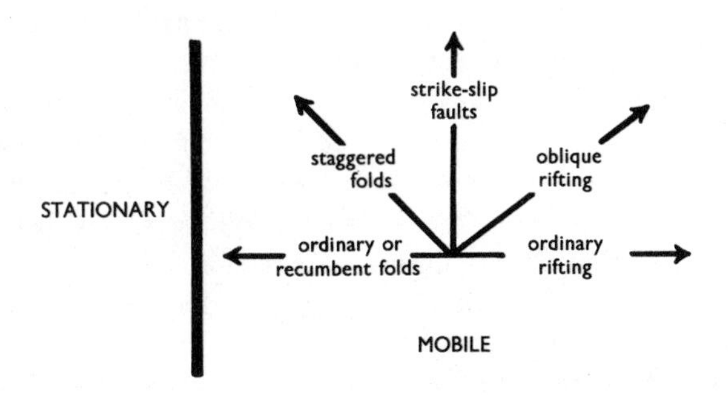


FIG. 49. Folding or rifting as a result of block movements with different directions.

arise, but the folds formed are specially large (overthrusts); if it moves at an oblique angle to the boundary, staggered folds are formed, which are closer and lower the more the direction of movement is parallel to the block edge. When there is precise parallelism, a slip plane results, with a strike-slip fault; finally, if the movement has a component directed away from the block boundary, we have either oblique or normal rifting which first appears as a trough fault. The ratio of normal to staggered folds can be shown very well with a table cloth, if we anchor the portion which represents the fixed block by means of weights, and push the other portion towards it.

Merely from these general considerations one can see that staggered folds must arise more often than normal ones, since the former represent the general case and the latter, the special case. The disposition of the fold ranges in nature appears to correspond to this. I would like to stress the point because geologists often recognise as properly homologous only such fold ranges as continue directly from one to the other, fold to fold, and from what has been said, this need not be so.

As shown in Figure 49, folds and rifts are but two different effects with the same cause, namely the displacement of block sections relative to each other, and they pass, via echelon folds and strike-slip faults, from one to the other in a continuous manner. It is therefore justifiable to consider the rift-fault process under the same heading.

The finest examples of such faults are provided by the East African rift valleys (grabens). They belong to a large fault system which can be traced northwards through the Red Sea, the Gulf of Aqaba and the Jordan Valley to the edge of the Taurus fold range (Fig. 50). According to recent investigations, these faults are continued southwards also, as far as Cape Province, but the finest examples of them

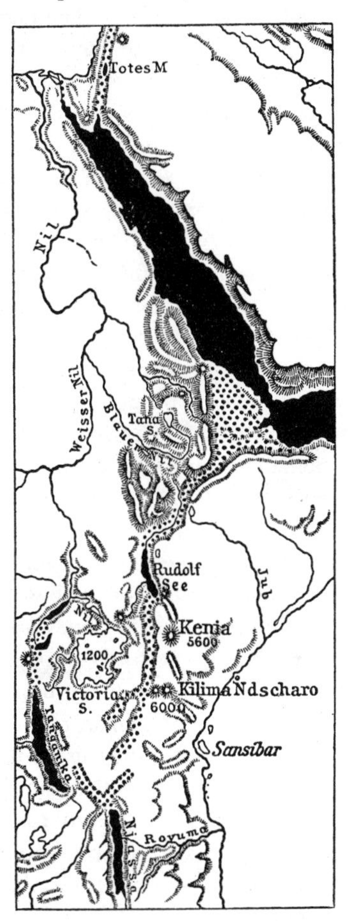


FIG. 50. The East African rift valleys, according to Supan. ∴ = rift valleys; ■ = portions of valley under water.

to be found in eastern Africa. Neumayr-Uhlig [183] describes them approximately as follows:

From the mouth of the Zambesi, a graben of this type, 50–80 km wide, runs north, taking in the Shiré River and Lake Nyasa, then turns northwest and disappears. In its place, close by and parallel to it, begins the graben of Lake Tanganyika, whose size is indicated by the fact that the depth of the lake is 1700–2700 m, but the height of the wall-like precipice 2000–2400, even 3000 m. In its northerly continuation, this rift valley includes the Russisi River, Lake Kivu, Lake Edward and Lake Albert. He goes on: "The margins of the valley appear ridged as if the earth had burst here with a certain upward movement of the fault margins as they were suddenly released. This peculiar protrusive shape of the edges of the plateau may well be connected with the fact that immediately east of the Lake Tanganyika precipice the sources of the Nile rise, while the lake itself drains into the Congo River." A third prominent rift valley begins east of Lake Victoria, taking in Lake Rudolf farther north and curving near Abyssinia towards the northeast, where it continues on to the Red Sea on one side and the Gulf of Aden on the other. In the coastal region and in the interior of what was formerly German East Africa, these faults mostly take the form of step faults whose eastern side is downthrown.

Of special interest is the large triangle, marked in Figure 50 by dots in the same way as the valley floors, which lies in the angle formed by Abyssinia and the Somali peninsula (between Ankober, Berbera and Massawa). This relatively flat, low-lying area is composed entirely of recent volcanic lavas. Most authors regard it as a vast broadening of the rift floor. This idea is suggested particularly by the course of the coastlines on either side of the Red Sea, whose otherwise accurate parallelism is spoilt by this projection; if one cuts this triangle out, the opposite corner of Arabia fits perfectly into the gap. It has already been mentioned that we are obviously dealing here with sial masses from the underside of the Abyssinian range, which onesidedly spread out northwards and emerged at the edge of the block. Perhaps the rift was already filled with basalt, so that the rising sial carried a top-layer of this material up with it. In any case, the considerable elevation above ocean level points to the presence of sial masses beneath the lava cover, unless the region possibly shows an appreciable positive gravitational anomaly.

The origin of these faults, which form a network in eastern Africa itself, should be dated in recent geological times. At several places, they intersect recent basalt lavas, at one point even Pliocene freshwater formations. In any event, therefore, they could not have arisen before the close of the Tertiary. On the other hand, they appear to have been already present in the Pleistocene, as may be concluded from the raised beaches—high-water markers for the lakes which lie on the rift floors and have no outflow. In the case of Lake Tanganyika, the so-called "relict fauna," obviously once marine, but later adapted to fresh water, also indicates that it has been in existence for a long time. The frequent earthquakes and intense volcanicity of the fault region, however, indicate that the process of rifting is still in progress today.

As for the mechanics of these rift valleys, the only new item which can assist the interpretation thereof is that they represent an early stage in a complete separation of the two block portions; it may be a question of recent, still incomplete rifting or else of an earlier attempt at rifting which died out because of reduction in the strength of the rift forces. According to our ideas, a complete separation would proceed approximately as follows: First, a gaping fissure would form only in the uppermost, more brittle layers, while the lower, plastic layers would stretch. Since the vertical ramps of the elevation in question would require too much compression strength of the rock, simultaneously with the rift or instead of it, there would be formed oblique slip planes, along which the marginal sections of the two parts of the block would drop into the rift, accompanied by many local earth tremors; this would all take place at the same rate as the rift opened up, so that the trench fault (graben) would always develop with only a moderate depth; its floor would consist of fault blocks of the same rock series, also cropping out sideways from the trench at the higher levels. At this stage, the trench fault is not yet isostatically compensated; according to E. Kohlschütter [184], this is also the case with a large number of the late East African rift valleys. There is now an uncompensated mass deficiency; therefore a corresponding gravitational anomaly is observed, and further, both edges of the rift are elevated to restore isostatic equilibrium, so that the impression arises that the rift passes through an anticlinal bulge along its length. The Black Forest and the Vosges Mountains on both sides of the Upper Rhine rift valley are well-known examples of this marginal ridge. If the rift finally becomes so deep that only the more plastic lower layers of sial still lie beneath it, they and their underlying viscid sima then rise, so that the previous mass deficient is made up and the graben is isostatically compensated as a whole. With further opening of the rift, its floor is first completely covered by the spreading of the plastic underlayers of the sial blocks, which are overlaid by fragments of the more brittle upper layers, till finally, when the rift is very wide, windows of sima appear. In the case of the huge Red Sea graben, development has gone so far that, as Triulzi and Hecker found, isostatic compensation has already taken place.

The fact that the uppermost layers of sial are essentially more brittle than the lower also explains the remarkable phenomenon that the edges of the blocks formerly in contact remain congruent when sial masses lie between them that appear to prohibit a smooth joining of the blocks. For example, the east coast of Madagascar and the west coast of India exhibit a remarkably straight fracture of the gneiss plateau on both sides, which hardly allows any other conclusion than that both parts were at one time directly connected . Yet the arcuate shelf of the Seychelles lies between them; this is clearly also composed of sial (the islands are granitic), and in our reconstruction would have to be moved into the gap. However, it seems more likely to me that we are concerned here only with the more plastic material of the deeper sial layers extracted during the rift process, which would be under the two block portions in our reconstruction; this naturally does not exclude the possibility that it can be capped by smaller surface fragments. The same holds good for the mid-Atlantic ridge and many other regions. It is important to bear this in mind, since otherwise it might appear puzzling in many areas that the outlines of the separated blocks are almost exactly congruent, while irregular masses of sial lie between them.

One may well attribute to this lateral extraction of the lower plastic layers of sial the fact that the margins of split-off continental blocks often drop down to the sea floor in the form of a series of step faults running parallel to the margin. They may often simulate an "anticlinal flexure" along the upper portion, the only portion investigated; that is, their surface forms an overhang. However, we cannot pursue these details any further here.

A special sort of force must operate at the margins of the plastic continental blocks when these are loaded by an inland ice sheet. If one applies stress-loading to a non-brittle cake, then in the attempt to reduce its thickness and expand radially, it will acquire radial cracks at the edges. This is the explanation for fjord formation, to be found with astonishing uniformity on all formerly glaciated coastlines (Scandinavia, Greenland, Labrador, the Pacific coast of North America north of the 48th parallel and of South America south of the 42nd, as well as South Island, New Zealand). Gregory [185] gave this explanation in a comprehensive and still greatly underestimated investigation of fault formation. Fjords are still often interpreted today as erosion valleys, but from my own observations in Greenland and Norway, I consider this to be wrong.

A large number of soundings have drawn attention to a peculiar phenomenon on the Atlantic continental margins, submarine continuations of river valleys. For example, the valley of the St. Lawrence River continues on through the coastal shelf to the ocean, and so does that of the Hudson River (detectable to the 1450-m depth contour). On the European side, the same applies to the estuary of the Tagus and particularly to the Fosse de Cap Breton, 17 km north of the Adour estuary. The most elegant case in point, however, is the Congo submarine canyon in the southern Atlantic (which can be followed to the 2000-m line). According to the usual interpretation, these canyons are submerged erosion valleys formed above the water line. This seems most improbable to me: first, because of the great depth of the channel; secondly, because of the widespread distribution (with sufficiently numerous soundings they will probably be found on every continental margin); and thirdly, because only a select group of river estuaries exhibit the phenomenon, while the intermediate estuaries do not. I think it more probable that this is a question of rifts in the continental margins which have been used by the rivers. Besides, in the case of the St. Lawrence, the rift-like nature of the river bed has been proved geologically; in the case of the Fosse de Cap Breton, which represents the innermost end of the open-book-shaped ocean rift of the Bay of Biscay, its whole position makes the rift explanation plausible.

The most interesting phenomenon of the continental margin, however, is that of the island arcs, which are a particular feature of the eastern coasts of Asia (Fig. 51). If we consider their distribution in the Pacific, we shall see a large-scale system. If, in particular, we conceive of New Zealand as a former island arc of Australia, the whole of the West Pacific coast is bordered by island arcs, while the east coast is not. In the area of North America, one could perhaps recognise the undeveloped beginnings of island-arc formation in the islands between 50 and 55° N, the coastal bulge near San Francisco and the separation of the Californian marginal chains. In the south, a claim might be put forward for western Antarctica as an island arc (probably a double one). By and large, however, the island-arc phenomenon signifies a displacement of the West Pacific continental masses, the direction being roughly west-northwest, or west considering the Pleistocene position of the poles; this direction also coincides with the long axis of the Pacific (South America–Japan) and with the

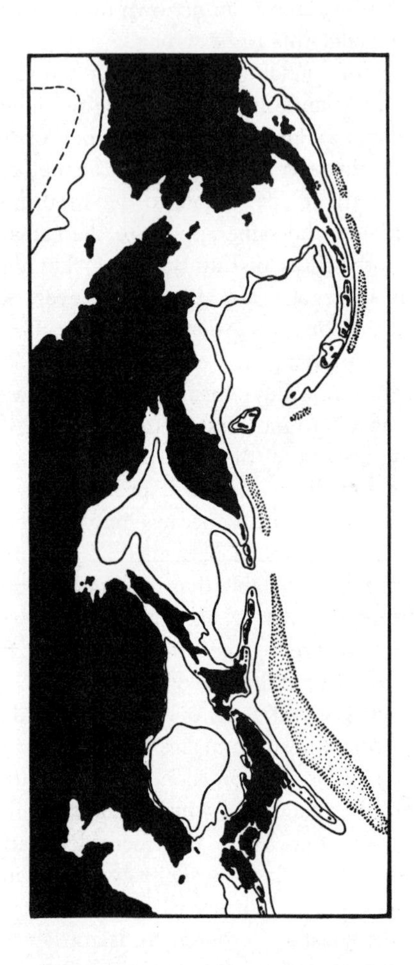

FIG. 51. Island arcs of northeastern Asia.
Depth contours 200 and 2000 m; ocean trenches dotted.

main line of the old Pacific island series (Hawaiian Islands, Marshall Islands, Society Islands, etc.) The ocean trenches, including the Tonga trench, are arranged as rifts perpendicular to this direction of displacement, and thus parallel to the island arcs. There is no doubt that all these things are causally connected with one another.

Quite similar island arcs are also found in the West Indies, and the South Shetland arc between Tierra del Fuego and Graham Land can also be claimed as a free island arc, though in a somewhat different sense of the expression.

The island arcs are echeloned uniformly in a very striking way. The Aleutians form a chain which, farther east in Alaska, is no longer a marginal chain, but an extension of inland ones. They terminate at Kamchatka, from which point the Kamchatka chain, an inland one up to this point, forms an arc with the Kurile Islands as the outermost chain. This in turn ends at Japan, giving place to the Sakhalin–Japan chain, till then an interior one. South of Japan, this arrangement can still be followed till, at the Sunda Islands, the situation becomes confusing. The Antilles also show exactly the same echelon formation. Obviously, this echelon formation of the island arcs is a direct result of the staggering of the former marginal mountain chains of the continents, and therefore goes back to the general law of echelon folding we mentioned previously. The remarkably equal lengths of the island arcs might perhaps already have been traced out tectonically in the outline of the marginal chains (the lengths are: Aleutians, 2900 km; Kamchatka–Kuriles, 2600; Sakhalin–Japan, 3000; Korea–Ryukyus, 2500; Formosa–Borneo, 2500; New Guinea–New Zealand, formerly 2700).[41]

Fujiwhara [195] has taken up the question of this echeloned formation, especially of the Japanese volcanic chains, and has tried to explain it by a rotation of the North Pacific sea floor in an anticlockwise direction (relative to the Asiatic block considered as a stationary reference frame). Since all motion is relative, one might conversely also consider this as a rotation of the surrounding land masses about the floor of the Pacific as reference, in a clockwise direction. This is of interest because the North Pole, until recent geological times, lay in the Pacific Ocean, so that such a rotation of the land masses in the past would correspond to a westerly drift. In fact, I think it very likely that the echeloned marginal chains of eastern Asia were established by such a drift of the continental blocks at the time when the pole still lay in the Pacific Ocean.

The remarkable conformity of the island arcs in their geological structure has already been mentioned: Their concave side always has a series of volcanoes, obviously a result of the pressure which was produced in forming the curve, and which extruded the sima in-

[41] The arcs of the West Indies, however, show a gradation: Lesser Antilles–southern Haiti–Jamaica–Mosquito Bank, 2600; Haiti–southern Cuba–Misteriosa Bank, 1900; Cuba 1100 km.

clusions. The convex side, on the other hand, has Tertiary sediments, which are mostly absent on the corresponding mainland shore. This would imply that the detachment of the arc took place only in very late geological times, and that it still formed the mainland margin at the time when the sediments were laid down. These Tertiary sediments show considerable overall disturbance, a consequence of the tension produced by curvature, leading to fissures and vertical faults. Honshu was split open at the Fossa Magna by the flexure, which was too strong to sustain. This outer edge of the island arc seems elevated, in spite of the fact that, otherwise, depression is generally associated with elongation; this probably implies a tilting movement of the arc, which one can imagine as having been caused by the end-points being entrained by the general westerly migration of the continental blocks, but the deep layers being held back by the sima. The ocean trenches, which usually are found at the outer edge of the arcs, seem to be associated with the same process. Attention has already been drawn to the fact that these trenches never occur on the freshly exposed sima surface between continent and arc, but always at the arc's outer margin only, that is, at the boundary of the old ocean floor. They appear here as rifts, one side of which is formed by the strongly-cooled old ocean floor, already solidified down to great depths, and the other side from the sialic material of the island arc. The formation of such a marginal rift between sial and sima would be very understandable in connection with this tilting motion to which we referred.

Again in Figure 51, the bulging outline of the continental margin behind the island arcs is also very striking. If, in particular, we consider the 200-m depth line besides the coastline itself, we see that the continental margin always forms the mirror image of an **S**, while the island arc which lies off the margin forms a simple convex curve. Figure 52B gives a diagrammatic illustration of this. The phenomenon shows up in all three of the arcs of Figure 51 in the same way, and it also occurs, for example, in the eastern Australian continental margin and its former island arc consisting of the southeastern offshoot of New Guinea and New Zealand. These bulging coastlines indicate a compression parallel to the coast, and thus also to the strike direction of the coastal mountain ranges. They should be considered as large-scale horizontal folds. We have here a subsidiary phenomenon in the vast compression experienced by the whole of eastern Asia in the northeast-southwest direction. If this wavy line of the eastern Asiatic

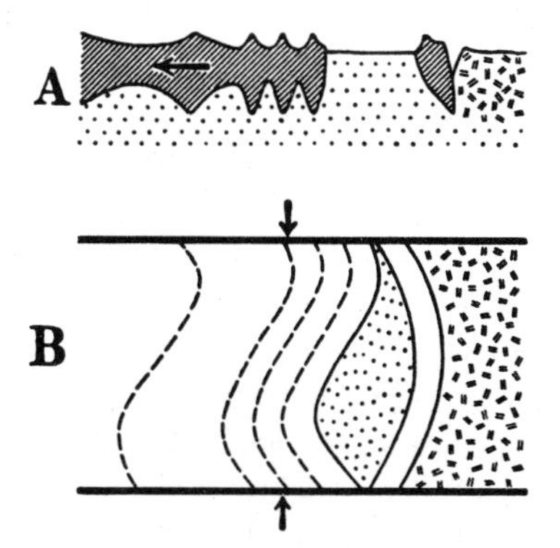


FIG. 52. Diagram of the origin of island arcs.
A: cross section; B: plan view. (The strongly cooled part of the sima is indicated by dashes.)

mainland coast is smoothed out, the distance between Indochina and the Bering Strait, now 9100 km, increases to 11,100 km.

According to our interpretation, the island arcs, and particularly the eastern Asiatic ones, are marginal chains which were detached from the continental masses when the latter drifted westwards and remained fast in the old sea floor, which was solidified to great depths. Between the arcs and the continental margin later, still-liquid areas of sea floor were exposed as windows.

This concept is quite different from that of F. von Richthofen, which admittedly starts out from very different assumptions [186]. He considered that the island arcs had arisen as the result of a tensile stress in the earth's crust originating in the Pacific. According to him, the island arcs, together with a broad area of the neighbouring mainland, which is also characterised by an arcuate coastline and elevations, formed a large-scale fault system. The region between island chain and mainland coast was the first "land step," sunk below sea level by a tilting movement on the west side, while the eastern edge projected as an island arc. Von Richthofen believed that two further such land steps were to be seen on the mainland; the subsidence of these, however, was of lesser degree. The regular arcuate form of these faults did indeed constitute a difficulty, but it was believed that this objection could be removed by reference to curved cracks in asphalt and other materials.

Although one must recognise that this theory was of historical service as the first deliberate breach with the dogma of a universally active "arching pressure" and the first advocacy of tensile stresses, it still does not take much to show that it does not do justice to our present-day data. In particular, the depth chart, though still incomplete for lack of soundings, constitutes decisive evidence that the link is broken between island arc and continental block.

When the movement of the continental blocks takes place, not perpendicular to the margin (as in eastern Asia), but parallel to it, then the marginal chains can be stripped off by strike-slip faulting, without any sima window appearing between them and the main block. This is basically the same set of phenomena as illustrated in Figure 49 for the interior of the continental block, transferred appropriately to the continental margin: If the block moves towards the sima, marginal folding results, in the form of either overthrusts or stepped folds, according to the direction of movement. If it moves away from the sea floor, the marginal chains split off. But if the movement is in shear, we have strike-slip faulting: the marginal chain slides. In this case also the marginal chain remains stuck in the solidified ocean floor. In our depth chart of Drake Strait (Fig. 26), one can see this process particularly well at the north end of Graham Land. In the same way, the southernmost chain of the Sunda Islands, Sumba–Timor–Ceram–Buru, formerly the southeastern continuation of the islands lying in front of Sumatra, slid past Java until caught by the advance of the Australia–New Guinea block.

Another example is California. The Baja California peninsula exhibits at its lateral projections entrainment phenomena (Fig. 53) which seem to indicate a forward thrust of the land mass to the south-southeast. The tip of the peninsula is thickened like an anvil by the frontal resistance of the sima, and the peninsula appears to be foreshortened overall, as shown by comparison with the cut-out area of the Gulf of California. According to Wittich [187], the northern part was only recently elevated above sea level to heights up to more than 1000 m, a sure sign of strong compression. The contours show almost beyond doubt that the tip was once to be found in the notch of the Mexican coast just below it. The geological chart shows in both areas "Post-Cambrian" intrusive rocks, the identity of which is, of course, not proved as yet.

Besides the foreshortening of the peninsula itself, however, there also appears to have been slip northwards, or, more correctly, the

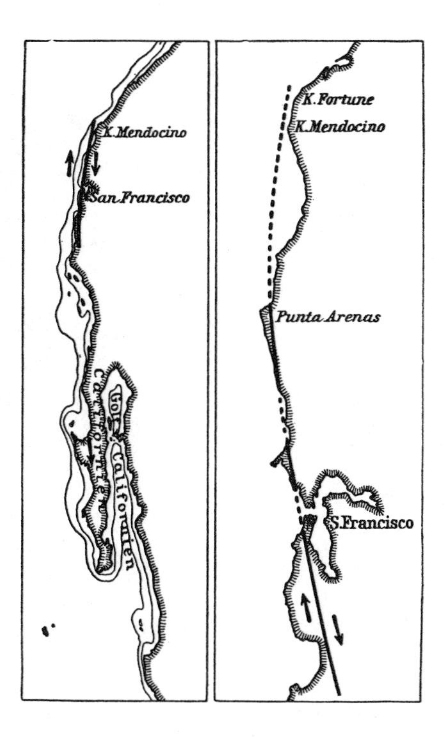


FIG. 53. California and the San Francisco earthquake fault.

peninsula seems to have remained behind during a southward thrust of the mainland *relative to the substratum*, in which the adjoining coastal chains to the north took part. This explains the large-scale bulging of the coastline near San Francisco, which must have suffered a compression. This idea is strikingly corroborated by the notorious San Francisco earthquake fault of the 18th of April 1906, indicated in our Figure 53, based on Rudzki [15] and Tams [188]. This is because the eastern section moved suddenly southwards, and the western section, northwards. As we would expect, the survey data showed that the extent of this sudden shift decreased with distance from the rift, and at large distances could not be detected. Naturally, the earth's crust was in slow continuous movement even before the fissure developed. Andrew C. Lawson [189] compared this movement between 1891 and 1906 with the direction of fissuring, and derived the result, valid for the "Point Arena group" of observations (Fig. 54), that an element

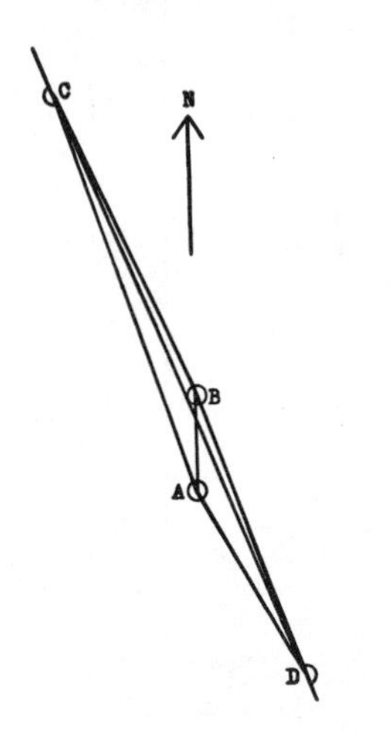


FIG. 54. The movement of a surface element intersected by the rift (according to Lawson).

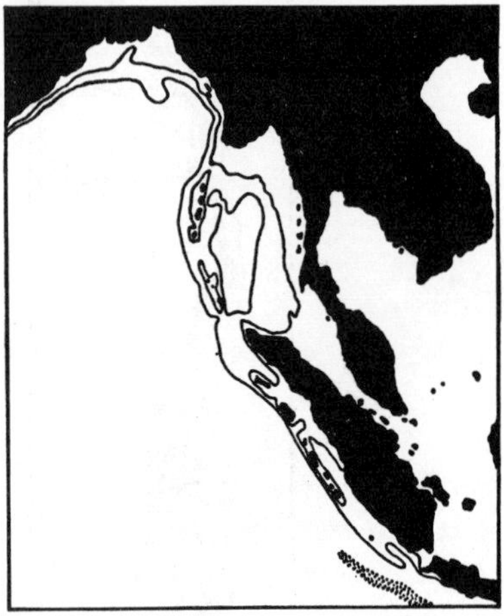

FIG. 55. Depth chart of Indochina.
Depth contours 200 and 2000 m; ocean trench is dotted.

of surface on the subsequent rift moved 0.7 m from A to B in the 15-year interval in question; it was then severed by the formation of the fault, whereupon the western half jumped 2.43 m to C, and the eastern 2.23 m to D. In the continuous movement between A and B, which must be thought of as relative to the main mass of the North American continent, the western continental margin is shown to have been held back towards the north by adhering to the Pacific sima. The fault only signifies a discontinuous release of stress, but the continental block as a whole did not shift.

In this connection we should also mention yet another very interesting part of the earth's crust, which is still little known, of course—the Indochina continental margin (Fig. 55). The point of chief interest here is the deep ocean basin north of Sumatra. The kink in the Malacca (Malay) Peninsula corresponds to the break-off of north Sumatra; but it is not possible to cover the window-like exposure of the deeper layers north of this island by straightening out the Malay Peninsula again. The island chain of the Andamans, to the west of the window, shows this to be so. We may perhaps assume here that the vast compression of the Himalayas put the Indochina chains in tension along their length, that under this stress the Sumatra chain was torn at the northern end of that island and that the northern part of the chain (Arakan) was, and still is, being pulled northwards like a rope's end into the great compression. On both sides of this grand-scale strike-slip fault, slip planes must have been formed. Interestingly enough, the outermost marginal chain, the Andaman and Nicobar Islands, remained stuck in the sima, and only the second chain underwent this remarkable displacement.

Finally, we should briefly consider the well-known distinction between "Pacific" and "Atlantic" types of coast. The latter coasts represent faults in a plateau, the former are characterised by the presence of marginal chains and ocean trenches in front of them. Among the Atlantic group we also count those of eastern Africa plus Madagascar, India, western and southern Australia and eastern Antarctica; the Pacific group includes the west coast of Indochina and the Sunda Archipelago, the east coast of Australia plus New Guinea and New Zealand, and western Antarctica. The West Indies plus the Antilles also have a Pacific structure. To the tectonic differences between these two types corresponds a different gravitational field strength, as shown by Meissner [190]. The Atlantic coasts are isostatically compensated, i.e., the floating continental blocks are in equilibrium here. The Pacific coasts largely show deviations from isostasy, however. It is also known that Atlantic coasts are relatively free of earthquakes and volcanoes, while Pacific coasts are well endowed with both. Becke has referred to the fact that, where an Atlantic-type coast produces a volcano, the lavas have systematic mineralogical differences from the Pacific lavas; they are denser and richer in iron, and apparently, therefore, originated at greater depths. According to our ideas, the "Atlantic" coasts are always those which were only formed since the Mesozoic, partly much later, by splitting of the continental blocks. The sea floor in front of them therefore represents a relatively freshly exposed deep layer, and must therefore be considered as relatively fluid. For this reason, there should be no surprise in finding these coasts isostatically compensated. Further, in drifting, the margins of the continents experience but little resistance because of this greater fluidity of the sima; they are therefore neither folded nor compressed, so that neither marginal chains nor volcanoes arise. In addition, no earthquakes are to be expected here, since the sima is fluid enough to facilitate all required movements without discontinuity, purely by plastic flow. The continents here behave like solid ice floes on liquid water, to express it in exaggerated form.

The earth's surface offers numerous indications of the fact that the essence of volcanicity is to be sought in a passive extrusion of sima inclusions through the sial crust. The island arcs are the most elegant examples of this. There, pressure must be produced on the concave inner side of the bend, and tearing on the convex outer side. Actually, their geological structure, as mentioned earlier, has a remarkable uniformity: The inner side is always marked by a series of volcanoes; the outer shows no volcanicity, but does exhibit strong cleavage and faulting. This ubiquitous arrangement of the volcanoes is so striking that it seems to me to be of the greatest significance with regard to the nature of volcanoes. Von Lozinski [191] writes: "In the Antilles, one can distinguish between a volcanic inner zone and two outer zones, of which the outermost consists of more recent beds and diminishes in height (Suess). The contrast between a highly volcanic inner zone and an outer zone with diminishing volcanicity obtains also in the Moluccas (Brouwer) and in Oceania (Arldt). The analogy with the arrangement of volcanic zones on the inner side of thrust zones, as in the Carpathian or Variscan hinterland, is obvious." The positions of Vesuvius, Etna and Stromboli correspond to this scheme; of the islands of the South Shetland arc between Tierra del Fuego and Graham Land, it is precisely the strongly curved central ridge of the South Sandwich Islands which is basaltic, and one of its volcanoes is still active. We have already referred to a specially interesting peculiarity of the Sunda Islands: Of the two southernmost island chains, only the singly curved northern one has volcanoes, but not the southern one (with Timor), which is the outer chain and therefore under tensile stress, and which, besides this, is curved in the opposite sense due to collision with the Australian shelf. At one place, however, near Wetar, the northern chain also is slightly indented because the southern chain (northeastern end of Timor) thrusts against it here; and just at this place on the northern chain, volcanicity has disappeared, though once active here also; this is obviously because the local curvature reverted. Brouwer also draws attention to the fact that the raised coral reefs, too, only occur where volcanicity is absent or extinct, which likewise indicates that these very regions are under compressive stress. The result, paradoxical at first sight, that volcanism ends where compression begins, finds an unforced explanation in the framework of our concepts.

It is not inconceivable that, in the earliest pre-geological times, the sial crust still covered the whole earth. It can only have had about a third of its present thickness and must have been covered by a "panthalassa" (universal sea), whose mean depth was calculated by A. Penck as 2.64 km, and which left little or nothing of the earth's surface exposed.

There are two factors which corroborate this concept—the development of life on the earth and the tectonics of the continental blocks.

Steinmann says [192]: "No one seriously doubts that the life in fresh water, dry land and air developed from marine life." Before the Silurian, we know of no air-breathing animals; the most ancient terrestrial plant remains date from the Upper Silurian in Gotland. According to Gothan [193], even from the Lower Devonian it is chiefly moss-like plants without proper leaves which are known: "Traces of real, spreading leaves are rare in the Lower Devonian. Almost all plant growth was small, herbaceous and not very sturdy." The Upper Devonian flora, on the other hand, was already similar to that of the Carboniferous "by virtue of the appearance of large, highly evolved, veined foliage, and of the successful division of labour in the plants with respect to the formation of supporting and assimilating organs. . . . The character of the Lower Devonian flora, its primitive organisation, small size, etc., suggests that the land flora derived from the water, a viewpoint already expressed by Potonié, Lignier, Arber and others. The advances observed for the Upper Devonian should be regarded as adaptation to the new way of life on land, in the atmosphere."

On the other hand, it appears as if, by smoothing out all the folds in the continental blocks, the sial crust would be sufficiently enlarged to cover the whole earth. Today, of course, the blocks plus their shelves cover only a third of the surface, but going back only as far as the Carboniferous we already find a considerable increase (about half the earth's surface covered). But the farther we go back in the earth's history, the more extensive are the folding processes. E. Kayser [34] writes: "It is of great significance, that the oldest Archæan rocks are strongly disturbed and folded everywhere on the earth. Only from the Algonkian onwards do we find here and there unfolded or weakly folded beds, as well as the folded. When we get to the post-Algonkian, we see that the extent and number of solid, unyielding masses becomes larger and larger, and correspondingly the extent of the foldable parts of the crust becomes more and more restricted. This holds especially for the Carboniferous-Permian compressions. In post-Palæozoic times the folding forces gradually decreased, but were intensified again in the Upper Jurassic and Cretaceous, reaching a new highpoint in the Upper Tertiary. It is, however, very significant that the region of propagation of even this latest large-scale mountain compression fell considerably short of the Carboniferous folding."

According to this, the idea that the sialsphere once enveloped the earth does not, in any case, conflict with the other viewpoints. This yielding and even plastic earth cover was now both split open and compressed by forces whose nature was discussed in Chapter 9. The origin and expansion of deep seas thus represent only one aspect of this process, whose other aspect consisted of folding. Biology appears to support the idea that the ocean was only formed during the course of the earth's history. Walther [194] writes: "General biological grounds, the stratigraphic arrangement of the present-day ocean fauna, as well as tectonic investigations, force us to believe that the ocean as a home of living creatures is not an original feature of the planet dating from the earliest times and that it was first formed when all portions of the present continents underwent tectonic fold movements and the relief of the earth's surface was fundamentally altered." The first cracks in the sialsphere, when the simasphere was first exposed, may have been similar to those which today form the East African rift valleys. They opened the wider, the greater the advances in sial folding. It was a process which we may roughly compare with the folding double of a round paper lantern: on one side, opening; on the other, compression. Very probably it was the area of the Pacific Ocean, generally regarded as very ancient, which was first denuded of its sial cover in this way. It is not inconceivable that the ancient folds in the gneiss massifs of Brazil, Africa, India and Australia represent the terrestrial equivalent of this opening up of the Pacific.

These compressions of the sialsphere must naturally have resulted in a thickening and outgrowth, while at the same time the ocean basins must have become larger. The flooding of the continental blocks must therefore have subsided more and more in the course of the earth's history; considered, of course, overall and apart from local variations. This rule is generally recognised. Consideration of our three reconstruction maps (Fig. 4) shows this very clearly.

It is important to note that the evolution of the sial crust must have been a unidirectional process even though the forces varied in direction. This is because the tensile stresses could not smooth out already-existing folds in a continental block, but could only tear it. An alternating process of compressive and tensile forces would therefore not be able to undo its own effects but would rather produce results always tending in the same direction: compression and disintegration. The sial cover becomes continuously smaller in area and thicker throughout the earth's history, but more and more subdivided. These processes supplement each other and are effects of the same cause. In Figure 56

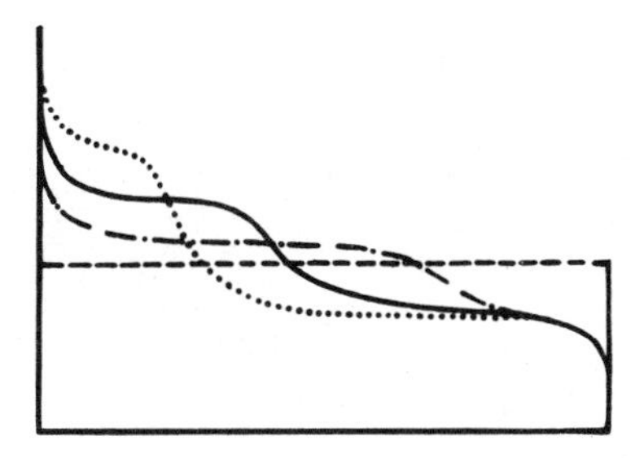

FIG. 56. Former and future hypsometric curves of the earth's surface.

.... = future; ——— = present; —·—· = past; – – = original state (also the mean crustal level).

we show the hypsometric curves which would result from these ideas, for the past, present and future. The mean crustal level is the same as the original surface of the uninterrupted sialsphere.

On the other hand, there is the possibility that the Pacific basin should be considered as the remains of the detachment of the moon, following Darwin's ideas, for this process would involve the loss of a portion of the sial crust of the earth. The only way to assess this, I believe, is to estimate the degree of compression-folding of the sial blocks. So far, however, there is no possibility of this.

CHAPTER 11

## *Supplementary Observations on the Ocean Floor*

MORPHOLOGICALLY, the ocean region contrasts as a whole with the continental blocks. The depths of the three great oceans are nevertheless not quite the same. Kossinna [29] calculated from the Groll depth charts that the mean depth of the Pacific Ocean is 4028 m, that of the Indian 3897 and that of the Atlantic 3332. A faithful picture of this relationship of depths is also given by the distribution of the ocean sediments (Fig. 57), which Krümmel once pointed out to me in person. The red clay and the radiolarian ooze, both genuine "abyssal"

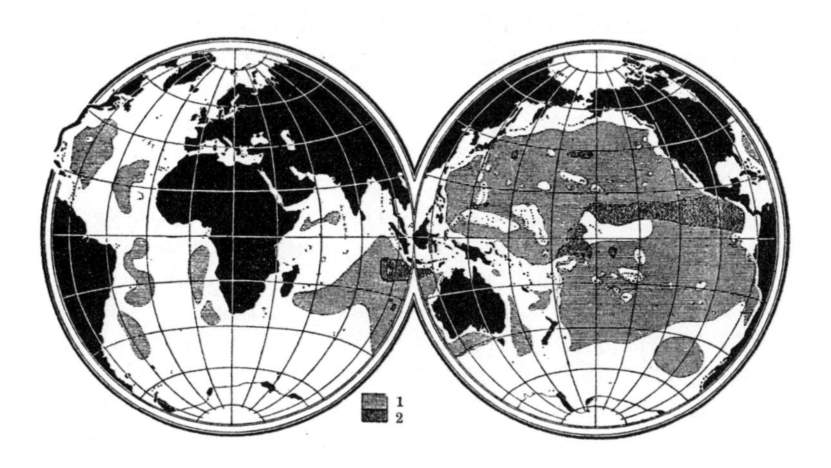


FIG. 57. Map of the ocean sediments, according to Krümmel [30].
1 = red clay; 2 = radiolarian ooze.

(deep-sea) sediments, are essentially confined to the Pacific and eastern Indian Oceans, while the Atlantic and western Indian Oceans are covered with "epilophic" sediments, whose higher lime content is causally connected with the oceans' lesser depth. These differences in depth are not accidental, but are systematically connected with the distinction between "Atlantic" and "Pacific" types of coast; this is best shown by the Indian Ocean, whose western half is Atlantic and whose eastern is Pacific in character; the eastern half is much deeper than the western. These matters therefore have considerable interest for drift theory, since a glance at the map shows that it is precisely the oldest sea floors which are the deepest, while those which were only relatively recently exposed are the least deep. Figure 57 shows in a surprising way the trail of the displacements, so to speak.

We have today no settled ideas about the cause of this difference in depths. It may consist in a difference of physical state, but also in dissimilarity of material. Physically, old and young sea floors can be differentiated by means of temperature or by the state of aggregation. If the density of the material is 2.9, and the cubical expansion coefficient of granite is taken as $26.9 \cdot 10^{-6}$, then the density for a temperature rise of 100°C would become 2.892. Two ocean floors, 100° different in temperature as far down as the 60-km depth line and in isostatic equilibrium with each other, must show a depth difference of 160 m, the warmer floor at the greater height.

On the other hand, it is not improbable that in the case of relatively freshly exposed sea floors, the crystalline solid covering is essentially thinner than in old ones, which could result in a difference in density and therefore in depth. If one considers all the ocean basins to have originated in the same way, the third possibility arises of assuming differences in material composition because of the large differences in date of origin; for, in the course of long geological periods, the magma can change due to progressive crystallisation or by some other effect, and presumably it has so changed. Finally, the sima areas may be covered to varying degrees by the remains of the flow of the underparts of the continental blocks, and by their marginal debris.

Our ideas on the material or materials of the sea floor are today, as we said, very much in a state of flux, so it is not worthwhile to cite all the views which have been expressed on the subject. I would therefore like to confine myself to a discussion of the most thoroughly investigated situation, that of the Atlantic Ocean, where, moreover, the wide mid-Atlantic ridge is a phenomenon which drift theory must cope with.

It has been known for a long time that the deep-sea floor often shows amazingly little variation in level over wide stretches. Such remarkably flat ocean regions have so far been found mainly by the closely spaced series of soundings made for cable-laying. Krümmel [30] mentioned that in the Pacific, over the 1540-km stretch between the Midway Islands and Guam, all 100 soundings lay between 5510 and 6277-m depth. On a 180-km stretch of this, the largest deviations from the mean depth of 5938 m were only +36 and −38 m for the 14 soundings taken. On another part, 550 m long, the deviations from a mean depth of 5790 m (37 soundings) did not exceed +103 and −112 m. Such closely spaced series of readings can now be more conveniently made by a ship in voyage with echo-sounding equipment. From the Atlantic region, the many depth profiles that the German "Meteor" expedition has obtained will soon provide further data. From the first echo-sounding profile of the North Atlantic, prepared by American researchers, I have shown [197] the western section, which cuts the ocean basin of the Sargasso just at its northernmost part, in Figure 58. This shows that between 58 and $47\frac{1}{2}°$ longitude

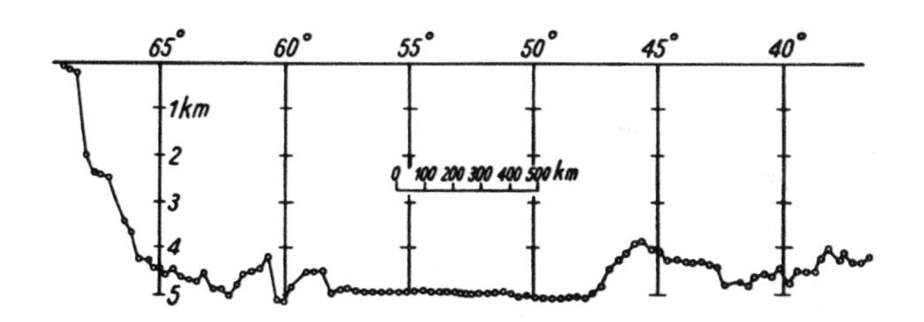


FIG. 58. Western section of the American echo-sounding profile through the North Atlantic, excluding the shelf.

(930 km) the largest deviations from the mean depth of 5132 m are −121 and −108 m. In some sections the constancy of depth is still more striking; for example, eight successive measurements, with 28 km between each two, lie between 2780 and 2790 fathoms (10 fathoms was the limit of precision of the measurements).

In contrast to this uniformity, the rest of the route shows a rough profile; it still forms part of the ocean floor, though less deep down.

I have deduced from this that, in the Sargasso Sea region, where the depth is so remarkably constant, the sima surface is exposed, while the other parts with rugged relief presumably correspond to a sial covering of varying and generally much lesser thickness than in the case of the continental blocks. If one makes the assumption, therefore, that everywhere below the 5000-m line the ocean floor corresponds roughly to exposed sima, then Figure 59 would show the surface distribution of sial and sima on the Atlantic floor.[42]

We now run into a certain difficulty. If we assume that these sial masses represent the remains of a strip which crumbled during the separation process, the strip would have to be extremely wide. Consider the route of the first transatlantic echo-sounding profile shown in Figure 59, for example. For the dispersed fragments, we would have to reckon on a strip 1300 km wide. In the South Atlantic we would, of course, obtain smaller values, since the mid-Atlantic ridge is narrower here and is bounded on both sides by ocean basins, not just on the west as in the route mentioned. More precise results will only be possible when the "Meteor" expedition soundings are made available; still we would arrive even here at a fragmented intermediate region of about 500 to 800 km. This is not absurd, but seems to me still excessive, since the congruence of the present-day block margins of South America and Africa is so striking here and seems to indicate that these margins were in fairly direct connection. Similar, though not very significant difficulties of this type are encountered at various other places in our reconstructions.

At the present time, it seems to me most likely that this small in-

[42] Gutenberg, making the same assumption that only the two materials sial and sima come under consideration, has expressed a different viewpoint, which he offers as a counter-theory to the drift concept—the "flow theory" [196]. He believes "that a single sial block floats on the sima, which is exposed only in the Pacific." He counts the Atlantic and Indian Ocean floors as a continental block and assumes that this block has here flattened out by half due to flow. But this cannot be correct. Even if we disregard the water loading, the depth of the Atlantic and Indian Oceans would then have to be only half that of the Pacific, and this difference would have to increase on isostatic grounds, due to the weight of the water. Gutenberg's view is therefore contradicted by the morphological similarity of the ocean floor as a whole and its contrast with the continental blocks; furthermore, if in our reconstructions the continents were moved towards each other to half their present-day separations, this would not meet the requirements of geology, biology and palæoclimatology; and finally, the remarkable congruence of the present-day margins of the continental blocks would remain a riddle. See above for further details.

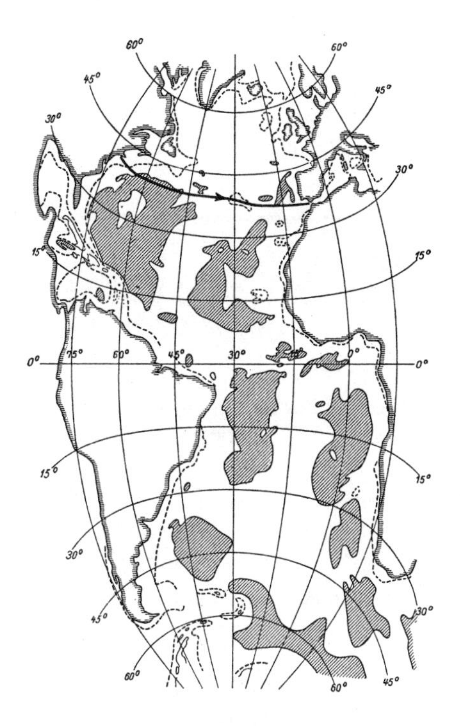

FIG. 59. The areas of the Atlantic floor below 5000-m depth.

congruity is caused by the fact that we have reckoned on only two layers—sial and sima—while in reality the situation is more complex. Instead of assuming this, let us assume something which appears to emerge more and more clearly from the latest geophysical investigations: that down to 30-km depth we normally have continental blocks of granite, below this to 60 km, basalt, and below this again, an ultrabasic rock (dunite). Then we arrive at an explanation which fits all the facts known today in a fully satisfactory way. The granite continental plateaux are in fact split apart, as assumed in drift theory, apart from certain deep, molten portions, and apart from the marginal fragments produced in the separations which today crown the mid-Atlantic ridge in the form of islands. If the basalt layer under the granite was really specially fluid as assumed, then, as the Atlantic rift opened wider progressively, this layer would have had to rise up here, subsequently flowing steadily out from both sides; it would first have formed the whole ocean floor, and would still today form the greater part of it. As the rift opened up progressively wider, the ability of even this material to flow must finally have become inadequate, and the underlying dunite must have been exposed as windows in the basalt (cf. Fig. 60). In the North Sea, where the separation of the

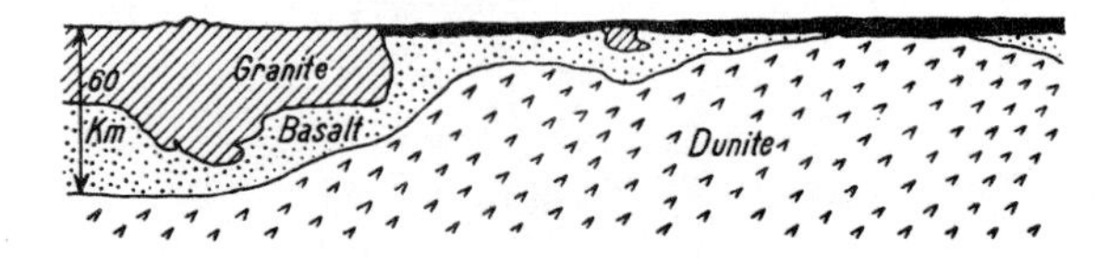


FIG. 60. Idealised section through a continental block and the sea floor.

blocks has still not progressed far, the floor would consist entirely of basalt, except for residual granite, and would still be of considerable thickness. In the vast area of the Pacific, however, correspondingly large areas of dunite would be exposed, while here too the flatter portions would still carry the basalt covering, perhaps topped here and there by granite remains.

Naturally, this picture is still entirely hypothetical. But I believe that my original concept of a fairly direct former connection between the continental blocks must be adhered to, according to all the geological, biological and palæoclimatological arguments. The recent investigations of the geophysicists in no way contradict it, as was shown; on the contrary, they appear to be able to remove the difficulty inherent in the fact that, between such blocks, which were obviously at one time directly joined, as their edges show, there are today irregular ridges on the ocean floors, as exemplified by the mid-Atlantic ridge. I shall not dispute the fact that, besides this, the continental blocks themselves may occasionally have been "drawn out" by flow, as Gutenberg would have it; we have made use of this idea at various places, in particular the Aegean Sea. However, flow proper should be restricted here too to the deeper layers, while the surface layers are broken up by faults.

There is no unanimity amongst present-day geophysicists on the question of how far basalt or dunite constitutes a material of the ocean floor, and therefore in what follows we would like to return to the simple distinction between sial and sima, for the sake of brevity.

If the sima is really a viscous material, it would be remarkable if its ability to flow were to be expressed only by yielding before the drifting sial blocks, and not by more independent flow movements. The map gives at certain places, by the distortion of apparently once straight island chains, a direct indication of such a more localised sima flow. In Figure 61 there are two examples of this: the Seychelles and the Fiji Islands. The half-moon shape of the Seychelles shelf, which carries the individual granitic islands, will fit neither Madagascar nor India, whose straight contours indicate rather a former direct connection. This suggests the interpretation that we are dealing here with fused sial masses that rose from beneath the block, were then carried off by the sima current and travelled a good part of the way towards India. This sima current, which Madagascar also followed, runs exactly in the track of India, perhaps produced by its drift or perhaps (just the reverse) producing it, as indicated by the breaking away of Ceylon. Movements in fluids, including viscous ones, are only rarely of so simple a kind that one can clearly separate cause from effect, and our knowledge of these matters is still much too defective. It is therefore foolish to demand of drift theory that it incorporate and explain every relative movement that arises. We consider these matters merely to elucidate flow phenomena in the sima; these phenomena are prominent particularly at the recurved ends of the shelves, which show that the movement of the sima current diminishes on either side of the centre-line Madagascar–India. We could also say: the current runs strongest in the freshly exposed sima, while the older sea floors move more slowly to the northwest and southeast from there. The lower illustration in Figure 61 shows, in the Fijis, a shape which is reminiscent of a two-armed spiral nebula and leads one to deduce a spiral flow. Its origin seems to be connected with the change of direction experienced by Australia when it broke its final link with Antarctica and, leaving behind the island arc of New Zealand, began its movement towards the northwest, still recognisable today. Presumably, before this coiling movement the Fijis formed a chain alongside and parallel to the Tonga ridge, and both together formed an outer island arc of the Australia–New Guinea block, which, like all East Asian arcs, adhered at the outer edge to the old sea floor, so that the inner edge became detached from the continental block; the inner chain was forced together into the form of a vortex by the retreat of the block. The New Hebrides and the Solomon Islands were probably

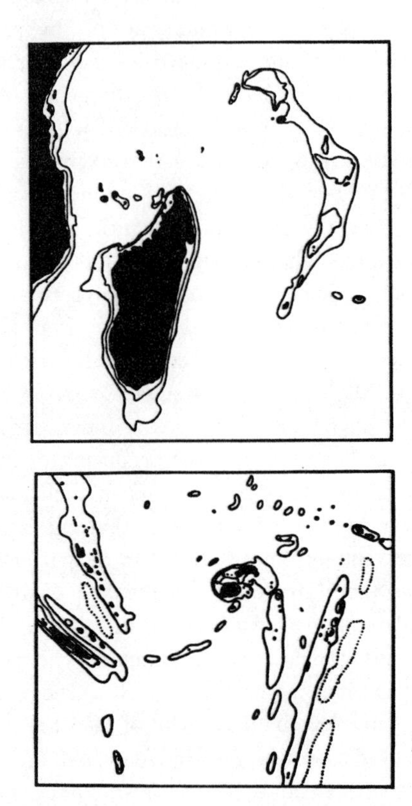

FIG. 61. ABOVE: Madagascar and the Seychelles Bank; BELOW: The Fiji Islands.

Depth contours 200 and 2000 m; deep-sea trenches shown dotted.

two other, echeloned island arcs, left behind en route.[43] New Britain, in the Bismarck Archipelago, remained adhering to New Guinea, and was dragged around, while on the other side of the great Australian block the spiral curve of the two southernmost chains of the Sundas indicates a vortical flow in the sima similar to that shown in the case of the Fijis.

From the observations to date, no conclusive picture of the nature of

[43] Hedley, from biological data, comes to the conclusion that New Guinea plus New Caledonia, the New Hebrides and the Solomon Islands form a unity.

ocean trenches[44] can be obtained. With few exceptions of perhaps different origin, they are always stretched in front of the outer (convex) sides of the island arcs where these border on *old* ocean floors, while on the inner side of the arcs, where the newly exposed ocean floor appears as windows, one never finds a trench. It therefore appears as though the old sea floor alone is capable of forming trenches because of its more intensive cooling and hardening. Perhaps one should regard the trenches as marginal rifts, one side formed from the sial of the island arc and the other from the sea-floor sima. The profile shown in Figure 62, in reality very flat, should not mislead one, because it is naturally very much evened out by gravity.

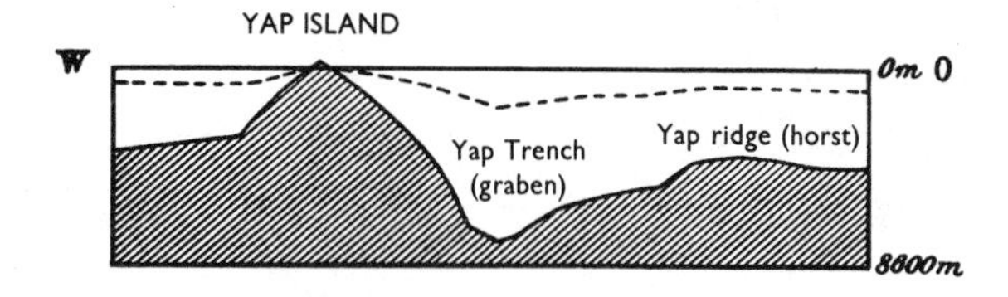


FIG. 62. Cross section through the Yap trench, vertical scale ×5 (according to G. Schott and P. Perlewitz).

The dashed line shows the true vertical scale.

In the case of the deep, rectangularly curved trench south and southeast of New Britain, the origin clearly derives from the powerful tug to the northwest experienced by the island when it adhered to New Guinea; the deeply embedded island block ploughed through the sima, which afterwards flowed into the gap and has still not quite filled it. Perhaps this is the case where we can render the most accurate account of the origin of an ocean trench.

It would appear that yet another kind of explanation could be given for the origin of the Atacama Trench west of Chile. Bearing in mind that, when these ranges were formed by resistance to the drift of the block, all layers beneath the ocean level were compressed downwards, we see that the neighbouring ocean floor must also have been drawn down with them.[45] There is still another reason for the depression

44 The term "deep-sea trough" is less happily chosen, since it includes the idea that these involve trough faults similar to those of the continental blocks.

45 The objection raised by Ampferer, A. Penck and others, that the westward movement of America must have piled up a sima range in front of the block

of the continental margin, namely the melting of downthrown mountain folds and—due to the westward drift of the block—the removal of the fused masses towards the east, where they in part come to the surface at the Abrolhos Bank, according to our assumptions. As a result, the continental margin must sink, entraining the neighbouring sima.

Naturally, these ideas require thorough, detailed testing. Of great importance here are the results of gravitational measurements. Hecker [198] had already found a large negative gravity anomaly over the Tonga Trench, but a positive anomaly on the nearby Tonga plateau. This was recently confirmed by Vening-Meinesz [39] for a large number of ocean trenches. From his publications, we

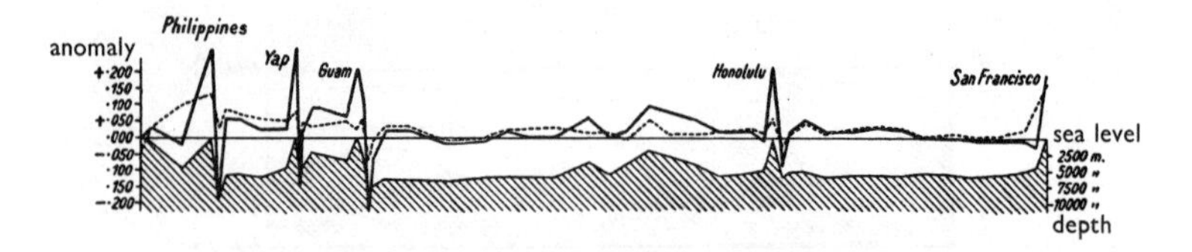


FIG. 63. Distribution of gravitational anomalies between the Philippines and San Francisco, according to Vening-Meinesz.

Dependent-variable scale at left; dotted line = data after applying the isostatic reduction; hatched area = sea floor; depth scale at right.

reproduce the gravity profile (Fig. 63) between the Philippines and San Francisco, on which the floor profile is also entered. Here four trenches were crossed and the gravity profile was everywhere alike: above the trench, a deficit; above the adjacent rise, an excess. This regularity seems to indicate that in the trench the isostatic adjustment by subsequent sima flow has not yet been achieved; the situation can perhaps be explained by assuming that the block exhibiting uplift is tilted (cf. Fig. 52). However, further research will be required before one can give a final verdict.

---

margin, is groundless if, as we must assume, all folds are formed under conservation of isostasy. The withdrawal of the sima cannot take place upwards because of its weight, but must only occur downwards, and backwards under the continental blocks, just like the flow of water when a floating body is towed slowly through it.

# *Appendix*

WHILE this book was in proof, a confirmation of the increase in distance between North America and Europe, claimed in Chapter 3, has been provided. We would not want to withhold this from the reader. F. B. Littell and J. C. Hammond have announced the results of the longitude-difference determinations carried out between North America and Europe in October and November of 1927. They have also compared these with the data obtained in 1913/1914.[46]

The differential longitude Washington–Paris in 1927 was

$$5\text{ h } 17\text{ m } 36.665\text{s} \pm 0.0019\text{ s},$$

but in 1913/1914

$$5\text{ h } 17\text{ m } 36.653\text{ s} \pm 0.0031\text{ s},$$
$$5\text{ h } 17\text{ m } 36.651\text{ s} \pm 0.003\text{ s}.$$

Of the two results for 1913/1914, the first relates to the measurements of the American observers and the second to those of the French.

From a comparison of these data, the Washington–Paris difference in longitude has increased in the course of 13–14 years by

$$0.013\text{ s} \pm \text{ca. } 0.003\text{ s},$$

or, in linear measure, by about

$$4.35\text{ m} \pm \text{ca. } 1.0\text{ m}.$$

This corresponds to an annual increase of distance of about

$$0.32\text{ m} \pm \text{ca. } 0.08\text{ m}.$$

The direction and amount of this change agree very well with the deductions on the basis of drift theory as given in Chapter 3.

[46] F. B. Littell and J. C. Hammond, "World Longitude Operation," *Astronomical Journal*, **38**, No. 908, p. 185, 14 Aug. 1928.

# *References*


[1] WEGENER, A., "Die Entstehung der Kontinente," *Petermanns Mitteilungen*, 1912, pp. 185–195, 253–256, 305–309.

[2] WEGENER, A., "Die Entstehung der Kontinente," *Geologische Rundschau*, **3**, No. 4, 1912, pp. 276–292.

[3] WEGENER, A., *Die Entstehung der Kontinente und Ozeane* (Sammlung Vieweg, No. 23, 94 pp.), Braunschweig, 1915; 2nd ed. (Die Wissenschaft, No. 66, 135 pp.), Braunschweig, 1920; 3rd ed., 1922.

[4] LÖFFELHOLZ VON COLBERG, Carl Freiherr, *Die Drehung der Erdkruste in geologischen Zeiträumen* (62 pp.), Munich, 1886. (2nd, much enlarged ed., 247 pp., Munich, 1895.)

[5] KREICHGAUER, D., *Die Äquatorfrage in der Geologie* (304 pp.), Steyl, 1902; 2nd ed., 1926.

[6] WETTSTEIN, H., *Die Strömungen der Festen, Flüssigen und Gasförmigen und ihre Bedeutung für Geologie, Astronomie, Klimatologie und Meteorologie* (406 pp.), Zurich, 1880.

[7] SCHWARZ, E. H. L., *Geological Journal*, 1912, pp. 294–299.

[8] PICKERING, *Journal of Geology*, **15**, No. 1, 1907; also *Gaea*, 43, 1907, p. 385.

[9] COXWORTHY, W. FRANKLIN, *The Electrical Condition, or How and Where Our Earth was Created*, London, W. J. S. Phillips, 1890(?).

[10] TAYLOR, F. B., "Bearing of the Tertiary Mountain Belt on the Origin of the Earth's Plan," *Bulletin of the Geological Society of America*, **21** (2), June 1910, pp. 179–226.

[11] ARLDT, T., *Handbuch der Paläogeographie*, Leipzig, 1917.

[12] SUESS, E., *Das Antlitz der Erde*, **1**, 1885.

[13] AMPFERER, "Über das Bewegungsbild von Faltengebirgen," *Jahrbuch der k. k. Geologischen Reichsanstalt*, **56**, Vienna, 1906, pp. 529–622.

[14] REYER, *Geologische Prinzipienfragen*, Leipzig, 1907.

[15] RUDZKI, M. P., *Physik der Erde*, Leipzig, 1911.

[16] ANDRÉE, K., *Über die Bedingungen der Gebirgsbildung*, Berlin, 1914.

[17] HEIM, A., "Bau der Schweizer Alpen," *Neujahrsblatt der Naturforschung-Gesellschaft*, Zurich, 1908, Part 110.

[18] STAUB, R., "Der Bau der Alpen," *Beiträge zur geologischen Karte der Schweiz*, N.F., No. 52, Bern, 1924.

[19] HENNIG, E., "Fragen zur Mechanik der Erdkrusten-Struktur," *Die Naturwissenschaften*, 1926, p. 452.

[20] ARGAND, E., "La tectonique de l'Asie," *Extrait du Compte-rendu du XIIIe Congrès géologique international 1922*, Liège, 1924.

[21] KOSSMAT, F., "Erörterungen zu A. Wegeners Theorie der Kontinentalverschiebungen," *Zeitschrift der Gesellschaft für Erdkunde zu Berlin*, 1921.

[22] DACQUÉ, E., *Grundlagen und Methoden der Paläogeographie*, Jena, 1915.

[23] GEER, G. DE, *Om Skandinaviens geografiska Utvekling efter Istiden, Stockholm*, 1896.

[24] KOBER, L., *Der Bau der Erde*, Berlin, 1921; *Gestaltungesgeschichte der Erde*, Berlin, 1925.

[25] STILLE, H., *Die Schrumpfung der Erde*, Berlin, 1922.

[26] NÖLCKE, F., *Geotektonische Hypothesen*, Berlin, 1924.

[27] WILLIS, B., "Principles of Palæogeography," *Science*, **31**, N.S., No. 790, 1910, pp. 241–260.

[28] WAGNER, H., *Lehrbuch der Geographie*, 1, Hanover, 1922.

[29] KOSSINNA, *Die Tiefen des Weltmeeres*, Veröffentlichungen des Instituts für Meereskunde, N.F.A., No. 9, Berlin, 1921.

[30] KRÜMMEL, *Handbuch der Ozeanographie*, Stuttgart, 1907.

[31] TRABERT, W., *Lehrbuch der kosmischen Physik*, Leipzig and Berlin, 1911.

[32] GROLL, M., *Tiefenkarten der Ozeane*, Veröffentlichungen des Instituts für Meereskunde, N.F.A., No. 2, Berlin, 1912.

[33] HEIM, A., *Untersuchungen über den Mechanismus der Gebirgsbildung*, Part 2, Basel, 1878.

[34] KAYSER, E., *Lehrbuch der allgemeinen Geologie*, 5th ed., Stuttgart 1918.

[35] SOERGEL, W., "Die atlantische 'Spalte': Kritische Bemerkungen zu A. Wegeners Theorie von der Kontinentalverschiebung," *Monatsberichte der Deutschen geologischen Gesellschaft*, **68**, 1916, pp. 200–239.

[36] DOUGLAS, G. V. and A. V., "Notes on the Interpretation of the Wegener Frequence Curve," *Geological Magazine*, **60**, No. 705, 1923.

[37] Wegener, A., "Der Boden des Atlantischen Ozeans," *Gerlands Beiträge zur Geophysik*, **17**, No. 3, 1927, pp. 311–321.

[38] Kossmat, F., "Die Beziehungen zwischen Schwereanomalien und Bau der Erdrinde," *Geologische Rundschau*, **12**, 1921, pp. 165–189.

[39] Vening-Meinesz, F. A., "Provisional Results of Determinations of Gravity, Made during the Voyage of Her Majesty's Submarine K XIII from Holland via Panama to Java," *Koninklijk Akademie van Wetenschappen te Amsterdam, Proc.*, Vol XXX, No. 7, 1927; "Gravity Survey by Submarine via Panama to Java," *Geographical Journal*, LXXI, No. 2, Feb. 1928. For the geological significance see A. Born, "Die Schwereverhältnisse auf dem Meere auf Grund der Pendelmessungen von Prof. Vening-Meinesz 1926," *Zeitschrift für Geophysik*, **3**, No. 8, 1927, p. 400.

[40] Schweydar, W., "Bemerkungen zu Wegeners Hypothese der Verschiebung der Kontinente," *Zeitschrift der Gesellschaft für Erdkunde zu Berlin*, 1921, pp. 120–125.

[41] Heiskanen, W., *Untersuchungen über Schwerkraft und Isostasie*, Veröffentlichungen der Finnischen Geodätischen Instituts, No. 4, Helsinki, 1924.

[42] Heiskanen, W., "Die Airysche isostatische Hypothese und Schweremessung," *Zeitschrift für Geophysik*, **1**, 1924/25, p. 225.

[43] Born, A., *Isostasie und Schweremessung*, Berlin, 1923.

[44] Gutenberg, B., *Der Aufbau der Erde*, Berlin, 1925.

[45] Gutenberg, B., *Lehrbuch der Geophysik*, Berlin, 1927/28, in process of publication.

[46] Tams, E., "Über die Fortpflanzungsgeschwindigkeit der seismischen Oberflächenwellen längs kontinentaler und ozeanischer Wege," *Centralblatt für Mineralogie, Geologie und Paläontologie*, 1921, pp. 44–52, 75–83.

[47] Angenheister, G., "Beobachtungen an pazifischen Beben," *Nachrichten der Königlichen Gesellschaft der Wissenschaften zu Göttingen, Math.-Phys. Klasse*, 1921, 34 pp.

[48] Visser, S. W., *On the Distribution of Earthquakes in the Netherlands East Indian Archipelago 1909/19*, Batavia, 1921.

[49] Wellman, H., *Über die Untersuchung der Perioden der Nachläuferwellen in Fernbebenregistrierungen auf Grund Hamburger und geeigneter Beobachtungen*, Diss., Hamburg, 1922.

[50] Wilde, H., *Proceedings of the Royal Society*, June 19, 1890 and Jan. 22, 1891.

[51] Rücker, A. W., "The Secondary Magnetic Field of the Earth," *Terrestrial Magnetism and Atmospheric Electricity*, **4**, Mar./Dec. 1899, pp. 113–129.

[52] Raclot, *Comptes Rendus de l'Académie des Sciences*, **164**, 150, 1917.

[53] JEFFREYS, H., "On the Earth's Thermal History and Some Related Geological Phenomena," *Gerlands Beiträge zur Geophysik*, **18**, 1927, pp. 1–29.

[54] DALY, R. A., *Our Mobile Earth*, London, 1926.

[55] MOHOROVIČIĆ, "Über Nahbeben und über die Konstitution des Erd- und Mondinnern," *Gerlands Beiträge zur Geophysik*, **17**, 1927, pp. 180–231.

[56] JOLY, J., *The Surface History of the Earth*, Oxford, 1925; and an article with the same title in *Gerlands Beiträge zur Geophysik*, **15**, 1926, pp. 189–200.

[57] HOLMES, A., "Contributions to the Theory of Magmatic Cycles," *Geological Magazine*, **63**, 1926, pp. 306–329; also: *Journal of Geology*, June/July 1926; "Oceanic Deeps and the Thickness of the Continents," *Nature*, Dec. 3, 1927.

[58] JOLY, J., and J. H. J. POOLE, "On the Nature and Origin of the Earth's Surface Structure," *Philosophical Magazine*, 1927, pp. 1233–1246.

[59] GUTENBERG, B., "Der Aufbau der Erdkruste," *Zeitschrift für Geophysik*, **3**, No. 7, 1927, p. 371.

[60] PREY, A., "Über Flutreibung und Kontinentalverschiebung," *Gerlands Beträge zur Geophysik*, **15**, No. 4, 1926, pp. 401–411.

[61] SCHWEYDAR, W., *Untersuchungen über die Gezeiten der festen Erde*, Veröffentlichungen der Preussichen Geodätischen Instituts, N.F., No. 54, Berlin, 1912.

[62] SCHWEYDAR, W., *Die Polbewegung in Beziehung zur Zähigkeit und zu einer hypothetischen Magmaschicht der Erde*, Veröffentlichungen der Preussischen Geodätischen Instituts, N.F., No. 79, Berlin, 1919.

[63] GREEN, W. L., "The Causes of the Pyramidal Form of the Outline of the Southern Extremities of the Great Continents and Peninsulas of the Globe," *Edinburgh New Philosophical Journal*, **6**, n.s., 1857; also: *Vestiges of the Molten Globe*, 1875.

[64] MEYERMANN, B., "Die Westdrift der Erdoberfläche, "*Zeitschrift für Geophysik*, **2**, No. 5, 1926, p. 204.

[65] MEYERMANN, B., "Die Zähigkeit des Magmas," *Zeitschrift für Geophysik*, **3**, No. 4, 1927, pp. 135–136.

[66] SCHULER, M., "Schwankungen in der Länge des Tages," *Zeitschrift für Geophysik*, **3**, Nos. 2/3, 1927, p. 71.

[67] DALY, R. A., "The Earth's Crust and Its Stability; Decrease of the Earth's Rotational Velocity and its Geological Effects," *American Journal of Science*, Vol. V, May 1923, pp. 349–377.

[68] AMPFERER, O., "Über Kontinentverschiebungen," *Die Naturwissenschaften*, **13**, 1925, p. 669.

[69] SCHWINNER, R., "Vulkanismus und Gebirgsbildung: Ein Versuch," *Zeitschrift für Vulkanologie*, **5**, 1919, pp. 175–230.

[70] KIRSCH, G., *Geologie und Radioaktivität*, Vienna and Berlin (Springer), 1928, pp. 115 *et seq.*

[71] PENCK, A., "Hebungen und Senkungen," *Himmel und Erde*, **25**, 1 and 2 (separate, no year).

[72] KEIDEL, J., "La Geología de las Sierras de la Provincia de Buenos Aires y sus Relaciones con las Montañas de Sud Africa y los Andes," *Annales del Ministerio de Agricultura de la Nación, Sección Geología, Mineralogía y Minería*, Vol. XI, No. 3, Buenos Aires, 1916.

[73] KEIDEL, H., "Über das Alter, die Verbreitung und die gegenseitigen Beziehungen der verschiedenen tektonischen Strukturen in den argentinischen Gebirgen," Étude faite à la XIIe Session du Congrès géologique international, reproduite du Compte-rendu, pp. 671–687 (separate, no year).

[74] BROUWER, H. A., "De alkaligesteenten van de Serra do Gericino ten Noordwesten van Rio de Janeiro en de overeenkomst der eruptiefgesteenten van Brasilië en Zuid-Afrika," *Koninklijk Akademie van Wetenschappen te Amsterdam*, 1921, Part 29, pp. 1005–1020.

[75] DU TOIT, A. L., "The Carboniferous Glaciation of South Africa," *Transactions of the Geological Society of South Africa*, **24**, 1921, pp. 188–227.

[76] LEMOINE, "Afrique occidentale," *Handbuch der regionalen Geologie*, VII, 6A, Part 14, Heidelberg, 1913, p. 57.

[77] MAACK, R., "Eine Forschungsreise über das Hochland von Minas Geraes zum Paranahyba," *Zeitschrift der Gesellschaft für Erdkunde zu Berlin*, 1926, pp. 310–323.

[78] DU TOIT, A. L., *A Geological Comparison of South America with South Africa. With a Palæontological Contribution by F. R. Cowper Reed*, Carnegie Institution of Washington, Publ., No. 381, Washington, 1927.

[79] PASSARGE, *Die Kalahari*, Berlin, 1904.

[80] WINDHAUSEN, A., "Ein Blick auf Schichtenfolge und Gebirgsbau im südlichen Patagonien," *Geologische Rundschau*, **12**, 1921, pp. 109–137.

[81] GAGEL, C., "Die mittelatlantischen Vulkaninseln," *Handbuch der regionalen Geologie*, VII, 10, Part 4, Heidelberg, 1910.

[82] KOSSMAT, "Die mediterranen Kettengebirge in ihrer Beziehung zum Gleichgewichtszustande der Erdrinde," *Abhandlungen der Mathematisch-Physischen Klasse der Sächsischen Akademie der Wissenschaften*, **38**, No. 2, Leipzig, 1921.

[83] ANDRÉE, K. "Verschiedene Beiträge zur Geologie Kanadas," *Schriften der Gesellschaft zur Beförderung der gesamten Naturwissenschaft zu Marburg*, **13**, 7, Marburg, 1914, pp. 437 *et seq.*

[84] TILMANN, N., "Die Struktur und tektonische Stellung der kanadischen Appalachen," *Sitzungsberichte der naturwissenschaftlichen Abteilung der Niederrheinischen Gesellschaft für Natur- und Heilkunde in Bonn*, 1916.

[85] LAUGE-KOCH, "Stratigraphy of Northwest Greenland," *Meddelelser fra Dansk Geologisk Forening*, **5**, No. 17, 1920 (78 pp.).

[86] MANTOVANI, R., "L'Antarctide," *Je m'instruis*, Sept. 19, 1909, pp. 595–597.

[87] LEMOINE, "Madagaskar," *Handbuch der regionalen Geologie*, VII, 4, Part 6, Heidelberg, 1911.

[88] KLEBELSBERG, R. VON, "Die Pamir-Expedition des Deutschen und Österreichischen Alpen-Vereins vom geologischen Standpunkt," *Zeitschrift des Deutschen und Österreichischen Alpen-Vereins*, 1914 (XLV), pp. 52–60.

[89] WILCKENS, O., "Die Geologie von Neuseeland," *Die Naturwissenschaften*, 1920, No. 41; also: *Geologische Rundschau*, **8**, 1917, pp. 143–161.

[90] BROUWER, H. A., "On the Crustal Movements in the Region of the Curving Rows of Islands in the Eastern Part of the East-Indian Archipelago," *Koninklijk Akademie van Wetenschappen te Amsterdam, Proc.*, **22**, Nos. 7 and 8, 1916; also: *Geologische Rundschau*, **8**, Nos. 5–8, 1917; *Nachrichten der Gesellschaft der Wissenschaften zu Göttingen*, 1920.

[91] MOLENGRAAFF, G. A. F., "The Coral Reef Problem and Isostasy," *Koninklijk Akademie van Wetenschappen*, 1916, p. 621 note.

[92] VAN VUUREN, L., *Het Gouvernement Celebes. Proeve eener Monographie*, 1, 1920 (see especially pp. 6–50).

[93] WING EASTON, "Het ontstaan van den maleischen Archipel, bezien in het licht van Wegener's hypothesen," *Tijdscrift van het Koninklijk Nederlandsch Aardrijkskundig Genootschap*, **38**, No. 4, July 1921, pp. 484–512; also: "On Some Extensions of Wegener's Hypothesis and Their Bearing upon the Meaning of the Terms Geosynclines and Isostasy," *Verhandelingen van het Geologisch-Mijnbouwkundig Genootschap voor Nederland en Kolonien*, Geolog. Ser., Part V, July 1921, pp. 113–133. (Proposes some modifications to drift theory, which are unsatisfactory in my opinion.)

[94] SMIT SIBINGA, G. L., "Wegener's Theorie en het ontstaan van den oostelijken O. J. Archipel," *Tijdschrift van het Koninklijk Nederlandsch Aardrijkskundig Genootschap*, 2nd Ser., Part XLIV, 1927, 5th ed.

[95] Escher, B. G., *Over Oorzaak en Verband der inwendige geologische Krachten*, Leyden, 1922.

[96] Wanner, "Zur Tektonik der Molukken," *Geologische Rundschau*, **12**, 1921, p. 160.

[97] Molengraaff, G. A. F., *De Geologie der Zeeën van Nederlandsch-Oost-Indië* (reprinted from *De Zeeën van Nederlandsch-Oost-Indië*, Leyden, 1921).

[98] Gagel, C., *Beiträge zur Geologie von Kaiser-Wilhemsland*, Beiträge zur geologische Erforschung der Deutschen Schutzgebiete, No. 4 (55 pp.) Berlin, 1912.

[99] Sapper, K., "Zur Kenntnis Neu-Pommerns und des Kaiser-Wilhelmsland," *Petermanns Mitteilungen*, **56**, 1910, pp. 89–193.

[100] Kühn, F., "Der sogenannte 'Südantillen-Bogen' und seine Beziehungen," *Zeitschrift der Gesellschaft für Erdkunde zu Berlin*, Nos. 8/10, 1920, pp. 249–262.

[101] Taylor, F. B., "Greater Asia and Isostasy," *American Journal of Science*, July 1926, pp. 47–67.

[102] Jeffreys, H., *The Earth: Its Origin, History and Physical Constitution*, Cambridge University Press, 1924.

[103] Cloos, H., "Geologische Beobachtungen in Südafrika. IV. Granite des Tafellandes und ihre Raumbildung," *Neues Jahrbuch für Mineralogie, Geologie und Paläontologie*, Beilage-Band XLII, pp. 420–456.

[104] Gutenberg, B., "Mechanik und Thermodynamik des Erdkörpers," in Müller-Pouillet, Vol. V, 1 (*Geophysik*), Braunschweig, 1928.

[105] Matley, C. A., "The Geology of the Cayman Islands (British West Indies)," *Quarterly Journal of the Geological Society*, LXXXII, Part 3, 1926, pp. 352–387.

[106] Hermann, F., "Paléogéographie et genèse penniques," *Eclogæ Geologicæ Helvetæ*, XIX, No. 3, 1925, pp. 604–618.

[107] Evans, J. W., "Regions of Tension," *Proceedings of the Geological Society*, LXXXI, Part 2, London, 1925, pp. lxxx–cxxii.

[108] Diener, "Die Grossformen der Erdoberflächen," *Mitteilungen der k. k. geologischen Gesellschaft, Wien*, **58**, 1915, pp. 329–349; "Die marinen Reiche der Triasperiode," *Denkschrift der Akademie der Wissenschaften, Wien, Math.-Naturw. Klasse*, 1915.

[109] Jaworski, "Das Alter des südatlantischen Beckens," *Geologische Rundschau*, 1921, pp. 60–74.

[110] Penck, A., "Wegeners Hypothese der kontinentales Verschiebungen," *Zeitschrift der Gesellschaft für Erdkunde zu Berlin*, 1921, pp. 110–120.

[111] Penck, W., "Zur Hypothese der Kontinentalverschiebung," *Zeitschrift der Gesellschaft für Erdkunde zu Berlin*, 1921, pp. 130–143.

[112] BROUWER, H. A., "On the Non-existence of Active Volcanoes between Pantar and Dammer (East Indian Archipelago), in Connection with the Tectonic Movements in this Region," *Koninklijk Akademie van Wetenschappen te Amsterdam, Proc.*, 21, Nos. 6 and 7, 1917.

[113] WASHINGTON, H. S., "Comagmatic Regions and the Wegener Hypothesis," *Journal of the Washington Academy of Sciences*, **13**, Sept. 1923, pp. 339–347.

[114] NÖLKE, F., "Physikalische Bedenken gegen A. Wegeners Hypothese der Entstehung der Kontinente und Ozeane," *Petermanns Mitteilungen*, 1922, p. 114.

[115] STROMER, *Geographische Zeitschrift*, 1920, p. 287 *et seq.*

[116] ÖKLAND, F., "Einige Argumente aus der Verbreitung der nordeuropäischen Fauna mit Bezug auf Wegeners Verschiebungstheorie," *Nyt Magsin för Naturvetenskap*, **65**, 1927, pp. 339–363.

[117] UBISCH, L. VON, "Wegeners Kontinentalverschiebungstheorie und die Tiergeographie," *Verhandlungen der Physikalisch-Medizinischen Gesellschaft zu Würzburg*, 1921.

[118] COLOSI, G., "La teoria della traslazione dei continenti e le dottrine biogeografiche," *L'Universo*, **6**, No. 3, March 1925. (Contains further references on biogeography.)

[119] ECKHARDT, W. R., "Die Beziehungen der afrikanischen Tierwelt zur südasiatischen," *Naturwissenschaftliche Wochenschrift*, No. 51, 1922.

[120] OSTERWALD, H., "Das Problem der Aalwanderungen in Lichte der Wegenerschen Verschiebungstheorie," *Umschau*, 1928, pp. 127–128.

[121] WEGENER, A., "Die geophysikalischen Grundlagen der Theorie der Kontinentverschiebung," *Scientia*, Feb. 1927.

[122] IHERING, H. VON, *Die Geschichte des Atlantischen Ozeans*, Jena, 1927.

[123] DE BEAUFORT, L. F., "De beteekenis van de theorie van Wegener voor de zoögeografie," *Handelingen van het XXe Nederlandsch Natuur- en Geneeskundig Congress*, April 14/16, April 1925, Groningen.

[124] HERGESELL, H., "Die Abkühlung der Erde und die gebirgsbildenden Kräfte," *Beiträge zur Geophysik*, **2**, 1895, p. 153.

[125] SEMPER, "Das paläothermale Problem, speziell die klimatischen Verhältnisse des Eozäns in Europa und den Polargebieten," *Zeitschrift der Deutschen Geologischen Gesellschaft*, **48**, 1896, pp. 261 *et seq.*

[126] SCHRÖTER, article "Geographie der Pflanzen" in the *Handwörterbuch der Naturwissenschaften.*

[127] Köppen, W., "Das Klima Patagoniens im Tertiär und Quartär," *Gerlands Beiträge zur Geophysik*, **17**, 3, 1927, pp. 391–394.

[128] Wegener, A., "Bemerkungen zu H. v. Iherings Kritik der Theorien der Kontinentverschiebungen und der Polwanderungen," *Zeitschrift für Geophysik*, **4**, No. 1, 1928, pp. 46–48.

[129] Klebelsberg, R. von, "Die marine Fauna der Ostrauer Schichten," *Jahrbuch der k. k. Geologischen Reichsanstalt*, **62**, 1912, pp. 461–556.

[130] Huus, J., "Über die Ausbreitungshindernisse der Meerestiefen und die geographische Verbreitung der Ascidien," *Nyt Magasin för Naturvetenskap*, **65**, 1927.

[131] Scharff, "Über die Beweisgründe für eine frühere Landbrücke zwischen Nordeuropa und Nordamerika," (*Proceedings of the Royal Irish Academy*, **28**, 1, 1909, pp. 1–28; from Arldt's paper, *Naturwissenschaftliche Rundschau*, 1910).

[132] Petersen, W., "Eupithecia fenestrata Mill, als Zeuge einer tertiären Landverbindung von Nord-Amerika mit Europa," *Beiträge zur Kunde Estlands*, **9**, 1922, pp. 4–5.

[133] Hoffmann, H., "Moderne Probleme der Tiergeographie," *Die Naturwissenschaften*, **13**, 1925, pp. 77–83.

[134] Ubisch, L. von, "Stimmen die Ergebnisse der Aalforschung mit Wegeners Theorie der Kontinentalverschiebung überein?" *Die Naturwissenschaften*, **12**, 1924, pp. 345–348.

[135] Ardlt, T., "Südatlantische Beziehungen," *Petermanns Mitteilungen*, **62**, 1916, pp. 41–46.

[136] Handlirsch, A., "Beiträge zur exakten Biologie," *Sitzungsberichte der Wiener Akademie der Wissenschaften, Math.-Naturw. Klasse*, **122**, 1, 1913.

[137] Kubart, B., "Bemerkungen zu Alfred Wegeners Verschiebungstheorie," *Arbeiten des phytopaläontologischen Laboratoriums der Universität Graz*, II, 1926.

[138] Sahni, B., "The Southern Fossil Floras: a Study in the Plant-Geography of the Past," *Proceedings of the 13th Indian Science Congress*, 1926.

[139] Wallace, A. R., *Die geographische Verbreitung der Tiere*, translated into German by Meyer (2 vols.) Dresden, 1876. [*The Geographical Distribution of Animals*.]

[140] Bresslau, E., article "Plathelminthes" in the *Handwörterbuch der Naturwissenschaften*, **7**, 993; also: Zschokke, *Zentralblatt für Bakteriologie und Parasitologie*, 1904, p. 36.

[141] Marshall, P., "New Zealand," *Handbuch der regionalen Geologie*, VII, 1, 1911.

[142] Bröndsted, H. V., "Sponges from New Zealand. Papers from Dr. Th. Mortensen's Pacific Expedition 1914/16," *Videnskabelige Meddelelser fra Dansk Naturhistoriske Foren*, **77**, pp. 435–483; **81**, pp. 295–331.

[143] Meyrick, E., "Wegener's Hypothesis and the Distribution of Micro-Lepidoptera," *Nature*, 125, pp. 834–835.

[144] Simroth, "Über das Problem früheren Landzusammenhangs auf der südlichen Erdhälfte," *Geographische Zeitschrift*, 1901, pp. 665–676.

[145] Andrée, "Das Problem der Permanenz der Ozeane und Kontinente," *Petermanns Mitteilungen*, **63**, 1917, p. 348.

[146] Arldt, T. "Die Frage der Permanenz der Kontinente und Ozeane," *Geographischer Anzeiger*, **19**, 1918, pp. 2–12.

[147] Griesebach, A., *Die Vegetation der Erde nach ihrer klimatischen Anordnung. Ein Abriss der vergleichenden Geographie der Pflanzen* (2 vols., 528 and 632 pp.) Leipzig, 1872.

[148] Drude, O., *Handbuch der Pflanzengeographie*, Stuttgart, 1890, p. 487.

[149] Ubisch, L. von, "Hermann v. Ihering's 'Geschichte des Atlantischen Ozeans,'" *Petermanns Mitteilungen*, 1927, pp. 206–207.

[150] Irmscher, E., "Pflanzenverbreitung und Entwicklung der Kontinente. Studien zur genetischen Pflanzengeographie," *Mitteilungen aus dem Institut für allgemeine Botanik in Hamburg*, **5**, 1922, pp. 15–235.

[151] Köppen, W., and A. Wegener, *Die Klimate der geologischen Vorzeit* (256 pp.) Berlin, 1924.

[152] Studt, W., *Die heutige und frühere Verbreitung der Koniferen und die Geschichte ihrer Arealgestaltung*, Diss., Hamburg, 1926.

[153] Koch, F., "Über die rezente und fossile Verbreitung der Koniferen im Lichte neuerer geologischer Theorien," *Mitteilungen der Deutschen Dendrologischen Gesellschaft*, No. 34, 1924.

[154] Michaelsen, W., "Die Verbreitung der Oligochäten im Lichte der Wegenerschen Theorie der Kontinentenverschiebung und andere Fragen zur Stammesgeschichte und Verbreitung dieser Tiergruppe," *Verhandlungen des Naturwissenschaftlichen Vereins zu Hamburg im 1921* (37 pp.), Hamburg, 1922.

[155] Svedelius, N., "On the Discontinuous Geographical Distribution of Some Tropical and Subtropical Marine Algæ," *Arkiv för Botanik* of the Kungliga Svenska Vetenkapsakademien, **19**, No. 3, 1924.

[156] Köppen, W., *Die Klimate der Erde. Grundriss der Klimakunde*, Berlin and Leipzig, 1923.

[157] Paschinger, V., "Die Schneegrenze in verschiedenen Klimaten," *Petermanns Mitteilungen*, 1912, Supplement No. 173.

[158] Köppen, W., "Die Lufttemperatur an der Schneegrenze," *Petermanns Mitteilungen* (separate, no year).

[159] Arldt, T., "Die Ursachen der Klimaschwankungen der Vorzeit, besonders der Eiszeiten," *Zeitschrift für Gletscherkunde*, **11**, 1918.

[160] Chamberlin, Rollin T., "Objections to Wegener's Theory," 1928; in [228].

[161] Reibisch, P., "Ein Gestaltungsprinzip der Erde," 27. *Jahresbericht des Vereins für Erdkunde zu Dresden*, 1901, pp. 105-124; Part II (contains only unimportant supplementary matter), *Mitteilungen des Vereins für Erdkunde zu Dresden*, **1**, 1905, pp. 39–53; III. "Die Eiszeiten," *ibid.*, **6**, 1907, pp. 58–75.

[162] Simroth, H., *Die Pendulationstheorie*, Leipzig, 1907.

[163] Schuchert, C., "The Hypothesis of Continental Displacement," 1928; in [228].

[164] Jacobitti, E., *Mobilità dell'Asse Terrestre, Studio Geologico*, Turin, 1912.

[165] Molengraaf, G. A. F., "The Glacial Origin of the Dwyka Conglomerate," *Transactions of the Geological Society of South Africa*, **4**, 1898, pp. 103–115.

[166] du Toit, A., "The Carboniferous Glaciation of South Africa," *Transactions of the Geological Society of South Africa*, **24**, 1921, pp. 188–227.

[167] Koken, *Indisches Perm und die permische Eiszeit*, publ. of the N. Jahrbuch für Mineralogie, 1907.

[168] Sayles, R. W., "The Squantum Tillite," *Bulletin of the Museum of Comparative Zoölogy at Harvard College*, **56**, No. 2 (Geol. Series, Vol. 10), 1914.

[169] Potonié, H., "Die Tropensumpfflachmoornatur der Moore des produktiven Karbons," *Jahrbuch der Königlichen Preussischen Geologischen Landesanstalt*, **30**, Part 1, No. 3, Berlin, 1909; *Die Entstehung der Steinkohle*, 5th ed., Berlin, 1910, p. 164.

[170] Rudzki, "L'Âge de la terre, "*Scientia*, **13**, No. xxviii, 2, 1913, pp. 161–173.

[171] Dacqué, E., section "Paläogeographie" in the *Enzyklopädie der Erdkunde*, edited by Kende, Leipzig and Vienna, 1926.

[172] *Danmark-Ekspeditionen til Grönlands Nordöstkyst* 1906/08 *under Ledelsen af L. Mylius-Erichsen*, 6 (Meddelelser om Grönland, 46), Copenhagen, 1917.

[173] Burmeister, F., "Die Verschiebung Grönlands nach den astronomischen Längenbestimmungen," *Petermanns Mitteilungen*, 1921, pp. 225–227.

[174] Jensen, P. F., "Ekspeditionen til Vestgrönland Sommeren 1922," *Meddelelser om Grönland*, lxiii, Copenhagen, 1923, pp. 205–283.

[175] Wegener, A., "Ekspeditionen til Vestgrönland Sommeren 1922" (P. F. Jensen, *Meddelelser om Grönland*, lxiii, Copenhagen, 1923, pp. 205–283), *Die Naturwissenschaften*, 1923, pp. 982–983.

[176] Stück, E., "Breiten- und Längenbestimmungen in Westgrönland im Sommer 1922," *Annalen der Hydrographie*, 1923, pp. 290–292.

[177] Galle, "Entfernen sich Europa und Nordamerika voneinander?" *Deutsche Revue*, Feb. 1916.

[178] "Jahresberichte des Preussischen Geodätischen Instituts" in *Vierteljahrsschrift der Astronomischen Gesellschaft*, **51**, 139; also: *Astronomical Journal*, Nos. 673/674.

[179] Wanach, B., "Ein Beitrag zur Frage der Kontinentalverschiebung," *Zeitschrift für Geophysik*, **2**, 1926, pp. 161–163.

[180] Poisson, P., *L'Observatoire de Tananarive*, Paris, 1924; Colin, P. E., *Comptes Rendus de l'Académie des Sciences*, 5 March 1894, p. 512; also: *La Géographie*, **45**, 1926, pp. 354–355, where the positions are also given.

[181] Günther, *Lehrbuch der Geophysik*, 1, Stuttgart, 1897, p. 278.

[182] Lambert, W. D., "The Latitude of Ukiah and the Motion of the Pole," *Journal of the Washington Academy of Sciences*, **12**, No. 2, 19 Jan. 1922.

[183] Neumayr-Uhlig, *Erdgeschichte*, 1: *Allgemeine Geologie*, 2nd ed., Leipzig and Vienna, 1897, p. 367.

[184] Kohlschütter, E., "Über den Bau der Erdkruste in Deutsch-Ostafrika," *Nachrichten der Königlichen Gesellschaft der Wissenschaften zu Göttingen, Math.-Phys. Klasse*, 1911.

[185] Gregory, J. W., *The Nature and Origin of Fjords* (542 pp.) London, 1913.

[186] Richthofen, F. von, "Über Gebirgskettungen in Ostasien. Geomorphologische Studien aus Ostasien, 4," *Sitzungsberichte der Königlichen Preussischen Akademie der Wissenschaften zu Berlin, Phys.-Math. Klasse*, **40**, 1903, pp. 867–891.

[187] Wittich, E., "Über Meeresschwankungen an der Küste von Kalifornien," *Zeitschrift der Deutschen Geologischen Gesellschaft*, **64**, 1912, Monatsbericht, No. 11, pp. 505–512; "La emersión moderna de la costa occidental de la Baja California," *Mémoires de la Société "Alzate,"* **35**, Mexico, 1920, pp. 121–144.

[188] Tams, "Die Entstehung des kalifornischen Erdbebens vom 18. April 1906," *Petermanns Mitteilungen*, **64**, 1918, p. 77.

[189] Lawson, A. C., "The Mobility of the Coast Ranges of California," *University of California Publications in Geology*, **12**, No. 7, 1921, pp. 431–473.

[190] Meissner, O., "Isostasie und Küstentypus," *Petermanns Mitteilungen*, **64**, 1918, p. 221.

[191] Lozinski, W. von, "Vulkanismus und Zusammenschub," *Geologische Rundschau*, 9, 1918, pp. 65–98.

[192] Steinmann, "Die kambrische Fauna im Rahmen der organischen Gesamtentwicklung," *Geologische Rundschau*, **1**, 1910, p. 69.

[193] Gothan, "Neues von den ältesten Landpflanzen," *Die Naturwissenschaften*, **9**, 1921, p. 553.

[194] Walther, J., "Über Entstehung und Besiedelung der Tiefseebecken," *Naturwissenschaftliche Wochenschrift*, N.F., Vol. 3, No. 46.

[195] Fujiwhara, S., "On the Echelon Structure of Japanese Volcanic Ranges and Its Significance from the Vertical Point of View," *Gerlands Beiträge zur Geophysik*, **16**, Nos. 1/2, 1927.

[196] Gutenberg, B. "Die Veränderungen der Erdkruste durch Fliessbewegungen der Kontinentalscholle," *Gerlands Beiträge zur Geophysik*, **16**, 1927, pp. 239–247; **18**, 1927, pp. 225–246.

[197] Wegener, A., "Der Boden des Atlantischen Ozeans," *Gerlands Beiträge zur Geophysik*, **17**, No. 3, 1917, pp. 311–321.

[198] Hecker, O., *Bestimmung der Schwerkraft auf dem Indischen und Grossen Ozean und an den Küsten*, Zentralbureau der Internationalen Erdmessung, N.F., No. 16, Berlin, 1908.

[199] Eötvös, *Verhandlungen der* 17. *Allgemeinen Konferenz der Internationalen Erdmessung*, Part 1, 1913, p. 111.

[200] Köppen, W., "Ursachen und Wirkungen der Kontinentenverschiebungen und Polwanderungen," *Petermanns Mitteilungen*, 1921, pp. 145–149, 191–194 (see esp. p. 149); "Über Änderungen der geographischen Breiten und des Klimas in geologischer Zeit," *Geografiska Annaler*, 1920, pp. 285-299; "Zur Paläoklimatologie," *Meteorologische Zeitschrift*, 1921, pp. 97–101 (with a different figure); "Über die Kräfte, welche die Kontinentenverschiebungen und Polwanderungen bewirken," *Geologische Rundschau*, **12**, 1922, pp. 314–320.

[201] Epstein, P. S., "Über die Polflucht der Kontinente," *Die Naturwissenschaften*, **9**, No. 25, pp. 499–502.

[202] Lambert, W. D., "Some Mechanical Curiosities Connected with the Earth's Field of Force," *American Journal of Science*, Vol. II, Sept. 1921, pp. 129–158.

[203] Berner, R., *Sur la grandeur de la force qui tendrait à rapprocher un continent de l'équateur*, thesis presented to the Faculté des Sciences of the University of Geneva, 1925.

[204] Wavre, R., "Sur la force qui tendrait à rapprocher un continent de l'équateur," *Archives des Sciences physiques et naturelles*, Aug. 1925.

[205] MÖLLER, M., *Kraftarten und Bewegungsformen*, Braunschweig, 1922.

[206] LELY, U. P., "Een Proef die de Krachten demonstreert, welke de Continentendrift kan veroorzaken," "*Physica*," *Nederlandsch Tijdschrift voor Natuurkunde*, **7**, 1927, pp. 278–281.

[207] MEYER, S., and SCHWEYDLER, E., *Radioaktivität*, 2nd ed., Leipzig, 1927, pp. 558 *et seq.*

[208] WANACH, B., "Eine fortschreitende Lagenänderung der Erdachse," *Zeitschrift für Geophysik*, **3**, Nos. 2/3, pp. 102–105.

[209] As yet unpublished: letter from Oberstleutnant Jensen by permission of Prof. Nörlund.

[210] WATERSCHOOT VAN DER GRACHT, W. A. J. M. VAN, "Remarks regarding the papers offered by the other contributors to the symposion," 1928; in [228].

[211] SCHIAPARELLI, *De la rotation de la terre sous l'influence des actions géologiques* (paper presented at the Pulkovo Observatory on the occasion of its 50th anniversary; 32 pp.) St. Petersburg, 1889.

[212] THOMPSON, Sir William, Report of Section of Mathematics and Physics, *Report of the British Association*, 1876, p. 11.

[213] FERRIÉ, G., "L'opération des longitudes mondiales (octobre/novembre 1926)," *Comptes Rendus de l'Académie des Sciences*, **186**, Paris, 5 March 1928.

[214] STAUB, R., "Das Bewegungsproblem in der modernen Geologie," inaugural lecture, Zurich, 1928.

[215] STAUB, R., *Der Bewegungsmechanismus der Erde*, Berlin, 1928.

[216] SLUYS, M., "Les périodes glaciaires dans le Bassin Congolais," *Compte Rendu du Congrès de Bordeaux 1923 de l'Association Française pour l'Avancement des Sciences*, 30 July 1923.

[217] WEGENER, A., "Two Notes concerning My Theory of Continental Drift," 1928; in [228].

[218] KÖPPEN, W., "Muss man neben der Kontinentenverschiebung noch eine Polwanderung in der Erdgeschichte annehmen?" *Petermanns Mitteilungen*, 1925, pp. 160–162.

[219] HEISKANEN, W., *Die Erddimensionen nach den europäischen Gradmessungen*, Veröffentlichungen des Finnischen Geodätischen Instituts, No. 6, Helsinki, 1926.

[220] SCHUMANN, R., "Über Erdschollen-Bewegung und Polhöhenschwankung," *Astronomische Nachrichten*, **227**, No. 5442, 1926, pp. 289–304.

[221] LAMBERT, W. D., "The Variation of Latitude," *Bulletin of the National Research Council*, **10**, Part 3, No. 53, Washington, 1925, pp. 43–45.

[222] Nansen, F., "The Earth's Crust, Its Surface Forms, and Isostatic Adjustment," *Det Norske Videnskaps-Akademi i Oslo, I. Mat.-Naturv. Klasse*, 1927, No. 12 (121 pp.) Oslo, 1928.

[223] Byerly, P., "The Montana Earthquake of June 28, 1925, G.M.C.T.," *Bulletin of the Seismological Society of America*, **16**, No. 4, Dec. 1926.

[224] Bowie, W., *Isostasy* (275 pp.), New York, 1927.

[225] Jaschnov, W. A., "Crustacea von Nowaja Zemlja" (offprint from the *Berichte des Wissenschaftlichen Meeresinstituts*, Lief. 12, Moscow, 1925; Russian with summary in German).

[226] Diener, C., *Grundzüge der Biostratigraphie*, Leipzig and Vienna, 1925.

[227] Ubisch, L. von, "Tiergeographie und Kontinentalverschiebung," *Zeitschrift für induktive Abstammungs- und Vererbungslehre*, **47**, 1928, pp. 159–179.

[228] *Theory of Continental Drift, a Symposium on the Origin and Movement of Land Masses Both Inter-continental and Intra-continental, as Proposed by Alfred Wegener*; *by W. A. J. M. van Waterschoot van der Gracht, Bailey Willis, Rollin T. Chamberlin, John Joly, G. A. F. Molengraaff, J. W. Gregory, Alfred Wegener, Charles Schuchert, Chester R. Longwell, Frank Bursley Taylor, William Bowie, David White, Joseph T. Singewald, Jr., and Edward W. Barry. Published by the American Association of Petroleum Geologists* (240 pp.), London, 1928.

[229] Brennecke, E., "Die Aufgaben und Arbeiten des Geodätischen Instituts in Potsdam in der Zeit nach dem Weltkriege," *Zeitschrift für Vermessungswesen*, Nos. 23 and 24, 1927.

# Index

# SOME DOVER SCIENCE BOOKS

CHANCE, LUCK AND STATISTICS: THE SCIENCE OF CHANCE,
*Horace C. Levinson*

Theory of probability and science of statistics in simple, non-technical language. Part I deals with theory of probability, covering odd superstitions in regard to "luck," the meaning of betting odds, the law of mathematical expectation, gambling, and applications in poker, roulette, lotteries, dice, bridge, and other games of chance. Part II discusses the misuse of statistics, the concept of statistical probabilities, normal and skew frequency distributions, and statistics applied to various fields—birth rates, stock speculation, insurance rates, advertising, etc. "Presented in an easy humorous style which I consider the best kind of expository writing," Prof. A. C. Cohen, Industry Quality Control. Enlarged revised edition. Formerly titled *The Science of Chance*. Preface and two new appendices by the author. Index. xiv + 365pp. 5⅜ x 8. Paperbound $2.00

BASIC ELECTRONICS,
*prepared by the U.S. Navy Training Publications Center*

A thorough and comprehensive manual on the fundamentals of electronics. Written clearly, it is equally useful for self-study or course work for those with a knowledge of the principles of basic electricity. Partial contents: Operating Principles of the Electron Tube; Introduction to Transistors; Power Supplies for Electronic Equipment; Tuned Circuits; Electron-Tube Amplifiers; Audio Power Amplifiers; Oscillators; Transmitters; Transmission Lines; Antennas and Propagation; Introduction to Computers; and related topics. Appendix. Index. Hundreds of illustrations and diagrams. vi + 471pp. 6½ x 9¼.
Paperbound $2.75

BASIC THEORY AND APPLICATION OF TRANSISTORS,
*prepared by the U.S. Department of the Army*

An introductory manual prepared for an army training program. One of the finest available surveys of theory and application of transistor design and operation. Minimal knowledge of physics and theory of electron tubes required. Suitable for textbook use, course supplement, or home study. Chapters: Introduction; fundamental theory of transistors; transistor amplifier fundamentals; parameters, equivalent circuits, and characteristic curves; bias stabilization; transistor analysis and comparison using characteristic curves and charts; audio amplifiers; tuned amplifiers; wide-band amplifiers; oscillators; pulse and switching circuits; modulation, mixing, and demodulation; and additional semiconductor devices. Unabridged, corrected edition. 240 schematic drawings, photographs, wiring diagrams, etc. 2 Appendices. Glossary. Index. 263pp. 6½ x 9¼. Paperbound $1.25

GUIDE TO THE LITERATURE OF MATHEMATICS AND PHYSICS,
*N. G. Parke III*

Over 5000 entries included under approximately 120 major subject headings of selected most important books, monographs, periodicals, articles in English, plus important works in German, French, Italian, Spanish, Russian (many recently available works). Covers every branch of physics, math, related engineering. Includes author, title, edition, publisher, place, date, number of volumes, number of pages. A 40-page introduction on the basic problems of research and study provides useful information on the organization and use of libraries, the psychology of learning, etc. This reference work will save you hours of time. 2nd revised edition. Indices of authors, subjects, 464pp. 5⅜ x 8.
Paperbound $2.75

THE RISE OF THE NEW PHYSICS (formerly THE DECLINE OF MECHANISM), *A. d'Abro*
This authoritative and comprehensive 2-volume exposition is unique in scientific publishing. Written for intelligent readers not familiar with higher mathematics, it is the only thorough explanation in non-technical language of modern mathematical-physical theory. Combining both history and exposition, it ranges from classical Newtonian concepts up through the electronic theories of Dirac and Heisenberg, the statistical mechanics of Fermi, and Einstein's relativity theories. "A must for anyone doing serious study in the physical sciences," *J. of Franklin Inst.* 97 illustrations. 991pp. 2 volumes.
Vol. 1 Paperbound $2.25, Vol. 2 Paperbound $2.25,
The set $4.50

THE STRANGE STORY OF THE QUANTUM, AN ACCOUNT FOR THE GENERAL READER OF THE GROWTH OF IDEAS UNDERLYING OUR PRESENT ATOMIC KNOWLEDGE, *B. Hoffmann*
Presents lucidly and expertly, with barest amount of mathematics, the problems and theories which led to modern quantum physics. Dr. Hoffmann begins with the closing years of the 19th century, when certain trifling discrepancies were noticed, and with illuminating analogies and examples takes you through the brilliant concepts of Planck, Einstein, Pauli, de Broglie, Bohr, Schroedinger, Heisenberg, Dirac, Sommerfeld, Feynman, etc. This edition includes a new, long postscript carrying the story through 1958. "Of the books attempting an account of the history and contents of our modern atomic physics which have come to my attention, this is the best," H. Margenau, Yale University, in *American Journal of Physics.* 32 tables and line illustrations. Index. 275pp. 5⅜ x 8.
Paperbound $1.75

GREAT IDEAS AND THEORIES OF MODERN COSMOLOGY, *Jagjit Singh*
The theories of Jeans, Eddington, Milne, Kant, Bondi, Gold, Newton, Einstein, Gamow, Hoyle, Dirac, Kuiper, Hubble, Weizsäcker and many others on such cosmological questions as the origin of the universe, space and time, planet formation, "continuous creation," the birth, life, and death of the stars, the origin of the galaxies, etc. By the author of the popular *Great Ideas of Modern Mathematics.* A gifted popularizer of science, he makes the most difficult abstractions crystal-clear even to the most non-mathematical reader. Index. xii + 276pp. 5⅜ x 8½. Paperbound $2.00

GREAT IDEAS OF MODERN MATHEMATICS: THEIR NATURE AND USE, *Jagjit Singh*
Reader with only high school math will understand main mathematical ideas of modern physics, astronomy, genetics, psychology, evolution, etc., better than many who use them as tools, but comprehend little of their basic structure. Author uses his wide knowledge of non-mathematical fields in brilliant exposition of differential equations, matrices, group theory, logic, statistics, problems of mathematical foundations, imaginary numbers, vectors, etc. Original publications, 2 appendices. 2 indexes. 65 illustr. 322pp. 5⅜ x 8. Paperbound $2.00

THE MATHEMATICS OF GREAT AMATEURS, *Julian L. Coolidge*
Great discoveries made by poets, theologians, philosophers, artists and other non-mathematicians: Omar Khayyam, Leonardo da Vinci, Albrecht Dürer, John Napier, Pascal, Diderot, Bolzano, etc. Surprising accounts of what can result from a non-professional preoccupation with the oldest of sciences. 56 figures. viii + 211pp. 5⅜ x 8½. Paperbound $1.50

COLLEGE ALGEBRA, *H. B. Fine*
Standard college text that gives a systematic and deductive structure to algebra; comprehensive, connected, with emphasis on theory. Discusses the commutative, associative, and distributive laws of number in unusual detail, and goes on with undetermined coefficients, quadratic equations, progressions, logarithms, permutations, probability, power series, and much more. Still most valuable elementary-intermediate text on the science and structure of algebra. Index. 1560 problems, all with answers. x + 631pp. 5⅜ x 8. Paperbound $2.75

HIGHER MATHEMATICS FOR STUDENTS OF CHEMISTRY AND PHYSICS, *J. W. Mellor*
Not abstract, but practical, building its problems out of familiar laboratory material, this covers differential calculus, coordinate, analytical geometry, functions, integral calculus, infinite series, numerical equations, differential equations, Fourier's theorem, probability, theory of errors, calculus of variations, determinants. "If the reader is not familiar with this book, it will repay him to examine it," *Chem. & Engineering News.* 800 problems. 189 figures. Bibliography. xxi + 641pp. 5⅜ x 8. Paperbound $2.50

TRIGONOMETRY REFRESHER FOR TECHNICAL MEN, *A. A. Klaf*
A modern question and answer text on plane and spherical trigonometry. Part I covers plane trigonometry: angles, quadrants, trigonometrical functions, graphical representation, interpolation, equations, logarithms, solution of triangles, slide rules, etc. Part II discusses applications to navigation, surveying, elasticity, architecture, and engineering. Small angles, periodic functions, vectors, polar coordinates, De Moivre's theorem, fully covered. Part III is devoted to spherical trigonometry and the solution of spherical triangles, with applications to terrestrial and astronomical problems. Special time-savers for numerical calculation. 913 questions answered for you! 1738 problems; answers to odd numbers. 494 figures. 14 pages of functions, formulae. Index. x + 629pp. 5⅜ x 8. Paperbound $2.00

CALCULUS REFRESHER FOR TECHNICAL MEN, *A. A. Klaf*
Not an ordinary textbook but a unique refresher for engineers, technicians, and students. An examination of the most important aspects of differential and integral calculus by means of 756 key questions. Part I covers simple differential calculus: constants, variables, functions, increments, derivatives, logarithms, curvature, etc. Part II treats fundamental concepts of integration: inspection, substitution, transformation, reduction, areas and volumes, mean value, successive and partial integration, double and triple integration. Stresses practical aspects! A 50 page section gives applications to civil and nautical engineering, electricity, stress and strain, elasticity, industrial engineering, and similar fields. 756 questions answered. 556 problems; solutions to odd numbers. 36 pages of constants, formulae. Index. v + 431pp. 5⅜ x 8. Paperbound $2.00

INTRODUCTION TO THE THEORY OF GROUPS OF FINITE ORDER, *R. Carmichael*
Examines fundamental theorems and their application. Beginning with sets, systems, permutations, etc., it progresses in easy stages through important types of groups: Abelian, prime power, permutation, etc. Except 1 chapter where matrices are desirable, no higher math needed. 783 exercises, problems. Index. xvi + 447pp. 5⅜ x 8. Paperbound $3.00

NUMERICAL SOLUTIONS OF DIFFERENTIAL EQUATIONS,
*H. Levy & E. A. Baggott*

Comprehensive collection of methods for solving ordinary differential equations of first and higher order. All must pass 2 requirements: easy to grasp and practical, more rapid than school methods. Partial contents: graphical integration of differential equations, graphical methods for detailed solution. Numerical solution. Simultaneous equations and equations of 2nd and higher orders. "Should be in the hands of all in research in applied mathematics, teaching," *Nature.* 21 figures. viii + 238pp. 5⅜ x 8. Paperbound $1.85

ELEMENTARY STATISTICS, WITH APPLICATIONS IN MEDICINE AND THE BIOLOGICAL SCIENCES, *F. E. Croxton*

A sound introduction to statistics for anyone in the physical sciences, assuming no prior acquaintance and requiring only a modest knowledge of math. All basic formulas carefully explained and illustrated; all necessary reference tables included. From basic terms and concepts, the study proceeds to frequency distribution, linear, non-linear, and multiple correlation, skewness, kurtosis, etc. A large section deals with reliability and significance of statistical methods. Containing concrete examples from medicine and biology, this book will prove unusually helpful to workers in those fields who increasingly must evaluate, check, and interpret statistics. Formerly titled "Elementary Statistics with Applications in Medicine." 101 charts. 57 tables. 14 appendices. Index. vi + 376pp. 5⅜ x 8. Paperbound $2.00

INTRODUCTION TO SYMBOLIC LOGIC,
*S. Langer*

No special knowledge of math required — probably the clearest book ever written on symbolic logic, suitable for the layman, general scientist, and philosopher. You start with simple symbols and advance to a knowledge of the Boole-Schroeder and Russell-Whitehead systems. Forms, logical structure, classes, the calculus of propositions, logic of the syllogism, etc. are all covered. "One of the clearest and simplest introductions," *Mathematics Gazette.* Second enlarged, revised edition. 368pp. 5⅜ x 8. Paperbound $2.00

A SHORT ACCOUNT OF THE HISTORY OF MATHEMATICS,
*W. W. R. Ball*

Most readable non-technical history of mathematics treats lives, discoveries of every important figure from Egyptian, Phoenician, mathematicians to late 19th century. Discusses schools of Ionia, Pythagoras, Athens, Cyzicus, Alexandria, Byzantium, systems of numeration; primitive arithmetic; Middle Ages, Renaissance, including Arabs, Bacon, Regiomontanus, Tartaglia, Cardan, Stevinus, Galileo, Kepler; modern mathematics of Descartes, Pascal, Wallis, Huygens, Newton, Leibnitz, d'Alembert, Euler, Lambert, Laplace, Legendre, Gauss, Hermite, Weierstrass, scores more. Index. 25 figures. 546pp. 5⅜ x 8. Paperbound $2.25

INTRODUCTION TO NONLINEAR DIFFERENTIAL AND INTEGRAL EQUATIONS,
*Harold T. Davis*

Aspects of the problem of nonlinear equations, transformations that lead to equations solvable by classical means, results in special cases, and useful generalizations. Thorough, but easily followed by mathematically sophisticated reader who knows little about non-linear equations. 137 problems for student to solve. xv + 566pp. 5⅜ x 8½. Paperbound $2.00

AN INTRODUCTION TO THE GEOMETRY OF N DIMENSIONS,
*D. H. Y. Sommerville*
An introduction presupposing no prior knowledge of the field, the only book in English devoted exclusively to higher dimensional geometry. Discusses fundamental ideas of incidence, parallelism, perpendicularity, angles between linear space; enumerative geometry; analytical geometry from projective and metric points of view; polytopes; elementary ideas in analysis situs; content of hyper-spacial figures. Bibliography. Index. 60 diagrams. 196pp. 5⅜ x 8.
Paperbound $1.50

ELEMENTARY CONCEPTS OF TOPOLOGY, *P. Alexandroff*
First English translation of the famous brief introduction to topology for the beginner or for the mathematician not undertaking extensive study. This unusually useful intuitive approach deals primarily with the concepts of complex, cycle, and homology, and is wholly consistent with current investigations. Ranges from basic concepts of set-theoretic topology to the concept of Betti groups. "Glowing example of harmony between intuition and thought," David Hilbert. Translated by A. E. Farley. Introduction by D. Hilbert. Index. 25 figures. 73pp. 5⅜ x 8.
Paperbound $1.00

ELEMENTS OF NON-EUCLIDEAN GEOMETRY,
*D. M. Y. Sommerville*
Unique in proceeding step-by-step, in the manner of traditional geometry. Enables the student with only a good knowledge of high school algebra and geometry to grasp elementary hyperbolic, elliptic, analytic non-Euclidean geometries; space curvature and its philosophical implications; theory of radical axes; homothetic centres and systems of circles; parataxy and parallelism; absolute measure; Gauss' proof of the defect area theorem; geodesic representation; much more, all with exceptional clarity. 126 problems at chapter endings provide progressive practice and familiarity. 133 figures. Index. xvi + 274pp. 5⅜ x 8.
Paperbound $2.00

INTRODUCTION TO THE THEORY OF NUMBERS, *L. E. Dickson*
Thorough, comprehensive approach with adequate coverage of classical literature, an introductory volume beginners can follow. Chapters on divisibility, congruences, quadratic residues & reciprocity. Diophantine equations, etc. Full treatment of binary quadratic forms without usual restriction to integral coefficients. Covers infinitude of primes, least residues. Fermat's theorem. Euler's phi function, Legendre's symbol, Gauss's lemma, automorphs, reduced forms, recent theorems of Thue & Siegel, many more. Much material not readily available elsewhere. 239 problems. Index. I figure. viii + 183pp. 5⅜ x 8.
Paperbound $1.75

MATHEMATICAL TABLES AND FORMULAS,
*compiled by Robert D. Carmichael and Edwin R. Smith*
Valuable collection for students, etc. Contains all tables necessary in college algebra and trigonometry, such as five-place common logarithms, logarithmic sines and tangents of small angles, logarithmic trigonometric functions, natural trigonometric functions, four-place antilogarithms, tables for changing from sexagesimal to circular and from circular to sexagesimal measure of angles, etc. Also many tables and formulas not ordinarily accessible, including powers, roots, and reciprocals, exponential and hyperbolic functions, ten-place logarithms of prime numbers, and formulas and theorems from analytical and elementary geometry and from calculus. Explanatory introduction. viii + 269pp. 5⅜ x 8½.
Paperbound $1.25

A SOURCE BOOK IN MATHEMATICS,
*D. E. Smith*
Great discoveries in math, from Renaissance to end of 19th century, in English translation. Read announcements by Dedekind, Gauss, Delamain, Pascal, Fermat, Newton, Abel, Lobachevsky, Bolyai, Riemann, De Moivre, Legendre, Laplace, others of discoveries about imaginary numbers, number congruence, slide rule, equations, symbolism, cubic algebraic equations, non-Euclidean forms of geometry, calculus, function theory, quaternions, etc. Succinct selections from 125 different treatises, articles, most unavailable elsewhere in English. Each article preceded by biographical introduction. Vol. I: Fields of Number, Algebra. Index. 32 illus. 338pp. 5⅜ x 8. Vol. II: Fields of Geometry, Probability, Calculus, Functions, Quaternions. 83 illus. 432pp. 5⅜ x 8.
Vol. 1 Paperbound $2.00, Vol. 2 Paperbound $2.00,
The set $4.00

FOUNDATIONS OF PHYSICS,
*R. B. Lindsay & H. Margenau*
Excellent bridge between semi-popular works & technical treatises. A discussion of methods of physical description, construction of theory; valuable for physicist with elementary calculus who is interested in ideas that give meaning to data, tools of modern physics. Contents include symbolism; mathematical equations; space & time foundations of mechanics; probability; physics & continua; electron theory; special & general relativity; quantum mechanics; causality. "Thorough and yet not overdetailed. Unreservedly recommended," *Nature* (London). Unabridged, corrected edition. List of recommended readings. 35 illustrations. xi + 537pp. 5⅜ x 8. Paperbound $3.00

FUNDAMENTAL FORMULAS OF PHYSICS,
*ed. by D. H. Menzel*
High useful, full, inexpensive reference and study text, ranging from simple to highly sophisticated operations. Mathematics integrated into text—each chapter stands as short textbook of field represented. Vol. 1: Statistics, Physical Constants, Special Theory of Relativity, Hydrodynamics, Aerodynamics, Boundary Value Problems in Math, Physics, Viscosity, Electromagnetic Theory, etc. Vol. 2: Sound, Acoustics, Geometrical Optics, Electron Optics, High-Energy Phenomena, Magnetism, Biophysics, much more. Index. Total of 800pp. 5⅜ x 8.
Vol. 1 Paperbound $2.25, Vol. 2 Paperbound $2.25,
The set $4.50

THEORETICAL PHYSICS,
*A. S. Kompaneyets*
One of the very few thorough studies of the subject in this price range. Provides advanced students with a comprehensive theoretical background. Especially strong on recent experimentation and developments in quantum theory. Contents: Mechanics (Generalized Coordinates, Lagrange's Equation, Collision of Particles, etc.), Electrodynamics (Vector Analysis, Maxwell's equations, Transmission of Signals, Theory of Relativity, etc.), Quantum Mechanics (the Inadequacy of Classical Mechanics, the Wave Equation, Motion in a Central Field, Quantum Theory of Radiation, Quantum Theories of Dispersion and Scattering, etc.), and Statistical Physics (Equilibrium Distribution of Molecules in an Ideal Gas, Boltzmann Statistics, Bose and Fermi Distribution. Thermodynamic Quantities, etc.). Revised to 1961. Translated by George Yankovsky, authorized by Kompaneyets. 137 exercises. 56 figures. 529pp. 5⅜ x 8½.
Paperbound $2.50

MATHEMATICAL PHYSICS, *D. H. Menzel*
Thorough one-volume treatment of the mathematical techniques vital for classical mechanics, electromagnetic theory, quantum theory, and relativity. Written by the Harvard Professor of Astrophysics for junior, senior, and graduate courses, it gives clear explanations of all those aspects of function theory, vectors, matrices, dyadics, tensors, partial differential equations, etc., necessary for the understanding of the various physical theories. Electron theory, relativity, and other topics seldom presented appear here in considerable detail. Scores of definition, conversion factors, dimensional constants, etc. "More detailed than normal for an advanced text . . . excellent set of sections on Dyadics, Matrices, and Tensors," *Journal of the Franklin Institute*. Index. 193 problems, with answers. x + 412pp. 5⅜ x 8. Paperbound $2.50

THE THEORY OF SOUND, *Lord Rayleigh*
Most vibrating systems likely to be encountered in practice can be tackled successfully by the methods set forth by the great Nobel laureate, Lord Rayleigh. Complete coverage of experimental, mathematical aspects of sound theory. Partial contents: Harmonic motions, vibrating systems in general, lateral vibrations of bars, curved plates or shells, applications of Laplace's functions to acoustical problems, fluid friction, plane vortex-sheet, vibrations of solid bodies, etc. This is the first inexpensive edition of this great reference and study work. Bibliography, Historical introduction by R. B. Lindsay. Total of 1040pp. 97 figures. 5⅜ x 8. Vol. 1 Paperbound $2.50, Vol. 2 Paperbound $2.50, The set $5.00

HYDRODYNAMICS, *Horace Lamb*
Internationally famous complete coverage of standard reference work on dynamics of liquids & gases. Fundamental theorems, equations, methods, solutions, background, for classical hydrodynamics. Chapters include Equations of Motion, Integration of Equations in Special Gases, Irrotational Motion, Motion of Liquid in 2 Dimensions, Motion of Solids through Liquid-Dynamical Theory, Vortex Motion, Tidal Waves, Surface Waves, Waves of Expansion, Viscosity, Rotating Masses of Liquids. Excellently planned, arranged; clear, lucid presentation. 6th enlarged, revised edition. Index. Over 900 footnotes, mostly bibliographical. 119 figures. xv + 738pp. 6⅛ x 9¼. Paperbound $4.00

DYNAMICAL THEORY OF GASES, *James Jeans*
Divided into mathematical and physical chapters for the convenience of those not expert in mathematics, this volume discusses the mathematical theory of gas in a steady state, thermodynamics, Boltzmann and Maxwell, kinetic theory, quantum theory, exponentials, etc. 4th enlarged edition, with new material on quantum theory, quantum dynamics, etc. Indexes. 28 figures. 444pp. 6⅛ x 9¼. Paperbound $2.75

THERMODYNAMICS, *Enrico Fermi*
Unabridged reproduction of 1937 edition. Elementary in treatment; remarkable for clarity, organization. Requires no knowledge of advanced math beyond calculus, only familiarity with fundamentals of thermometry, calorimetry. Partial Contents: Thermodynamic systems; First & Second laws of thermodynamics; Entropy; Thermodynamic potentials: phase rule, reversible electric cell; Gaseous reactions: van't Hoff reaction box, principle of LeChatelier; Thermodynamics of dilute solutions: osmotic & vapor pressures, boiling & freezing points; Entropy constant. Index. 25 problems. 24 illustrations. x + 160pp. 5⅜ x 8. Paperbound $1.75

CELESTIAL OBJECTS FOR COMMON TELESCOPES,
*Rev. T. W. Webb*
Classic handbook for the use and pleasure of the amateur astronomer. Of inestimable aid in locating and identifying thousands of celestial objects. Vol I, The Solar System: discussions of the principle and operation of the telescope, procedures of observations and telescope-photography, spectroscopy, etc., precise location information of sun, moon, planets, meteors. Vol. II, The Stars: alphabetical listing of constellations, information on double stars, clusters, stars with unusual spectra, variables, and nebulae, etc. Nearly 4,000 objects noted. Edited and extensively revised by Margaret W. Mayall, director of the American Assn. of Variable Star Observers. New Index by Mrs. Mayall giving the location of all objects mentioned in the text for Epoch 2000. New Precession Table added. New appendices on the planetary satellites, constellation names and abbreviations, and solar system data. Total of 46 illustrations. Total of xxxix + 606pp. 5⅜ x 8. Vol. 1 Paperbound $2.25, Vol. 2 Paperbound $2.25
The set $4.50

PLANETARY THEORY,
*E. W. Brown and C. A. Shook*
Provides a clear presentation of basic methods for calculating planetary orbits for today's astronomer. Begins with a careful exposition of specialized mathematical topics essential for handling perturbation theory and then goes on to indicate how most of the previous methods reduce ultimately to two general calculation methods: obtaining expressions either for the coordinates of planetary positions or for the elements which determine the perturbed paths. An example of each is given and worked in detail. Corrected edition. Preface. Appendix. Index. xii + 302pp. 5⅜ x 8½. Paperbound $2.25

STAR NAMES AND THEIR MEANINGS,
*Richard Hinckley Allen*
An unusual book documenting the various attributions of names to the individual stars over the centuries. Here is a treasure-house of information on a topic not normally delved into even by professional astronomers; provides a fascinating background to the stars in folk-lore, literary references, ancient writings, star catalogs and maps over the centuries. Constellation-by-constellation analysis covers hundreds of stars and other asterisms, including the Pleiades, Hyades, Andromedan Nebula, etc. Introduction. Indices. List of authors and authorities. xx + 563pp. 5⅜ x 8½. Paperbound $2.50

A SHORT HISTORY OF ASTRONOMY, *A. Berry*
Popular standard work for over 50 years, this thorough and accurate volume covers the science from primitive times to the end of the 19th century. After the Greeks and the Middle Ages, individual chapters analyze Copernicus, Brahe, Galileo, Kepler, and Newton, and the mixed reception of their discoveries. Post-Newtonian achievements are then discussed in unusual detail: Halley, Bradley, Lagrange, Laplace, Herschel, Bessel, etc. 2 Indexes. 104 illustrations, 9 portraits. xxxi + 440pp. 5⅜ x 8. Paperbound $2.75

SOME THEORY OF SAMPLING, *W. E. Deming*
The purpose of this book is to make sampling techniques understandable to and useable by social scientists, industrial managers, and natural scientists who are finding statistics increasingly part of their work. Over 200 exercises, plus dozens of actual applications. 61 tables. 90 figs. xix + 602pp. 5⅜ x 8½.
Paperbound $3.50

PRINCIPLES OF STRATIGRAPHY,
*A. W. Grabau*
Classic of 20th century geology, unmatched in scope and comprehensiveness. Nearly 600 pages cover the structure and origins of every kind of sedimentary, hydrogenic, oceanic, pyroclastic, atmoclastic, hydroclastic, marine hydroclastic, and bioclastic rock; metamorphism; erosion; etc. Includes also the constitution of the atmosphere; morphology of oceans, rivers, glaciers; volcanic activities; faults and earthquakes; and fundamental principles of paleontology (nearly 200 pages). New introduction by Prof. M. Kay, Columbia U. 1277 bibliographical entries. 264 diagrams. Tables, maps, etc. Two volume set. Total of xxxii + 1185pp. 5⅜ x 8. Vol. 1 Paperbound $2.50, Vol. 2 Paperbound $2.50,
The set $5.00

SNOW CRYSTALS, *W. A. Bentley and W. J. Humphreys*
Over 200 pages of Bentley's famous microphotographs of snow flakes—the product of painstaking, methodical work at his Jericho, Vermont studio. The pictures, which also include plates of frost, glaze and dew on vegetation, spider webs, windowpanes; sleet; graupel or soft hail, were chosen both for their scientific interest and their aesthetic qualities. The wonder of nature's diversity is exhibited in the intricate, beautiful patterns of the snow flakes. Introductory text by W. J. Humphreys. Selected bibliography. 2,453 illustrations. 224pp. 8 x 10¼. Paperbound $3.25

THE BIRTH AND DEVELOPMENT OF THE GEOLOGICAL SCIENCES,
*F. D. Adams*
Most thorough history of the earth sciences ever written. Geological thought from earliest times to the end of the 19th century, covering over 300 early thinkers & systems: fossils & their explanation, vulcanists vs. neptunists, figured stones & paleontology, generation of stones, dozens of similar topics. 91 illustrations, including medieval, renaissance woodcuts, etc. Index. 632 footnotes, mostly bibliographical. 511pp. 5⅜ x 8. Paperbound $2.75

ORGANIC CHEMISTRY, *F. C. Whitmore*
The entire subject of organic chemistry for the practicing chemist and the advanced student. Storehouse of facts, theories, processes found elsewhere only in specialized journals. Covers aliphatic compounds (500 pages on the properties and synthetic preparation of hydrocarbons, halides, proteins, ketones, etc.), alicyclic compounds, aromatic compounds, heterocyclic compounds, organophosphorus and organometallic compounds. Methods of synthetic preparation analyzed critically throughout. Includes much of biochemical interest. "The scope of this volume is astonishing," *Industrial and Engineering Chemistry*. 12,000-reference index. 2387-item bibliography. Total of x + 1005pp. 5⅜ x 8. Two volume set, paperbound $4.50

THE PHASE RULE AND ITS APPLICATION,
*Alexander Findlay*
Covering chemical phenomena of 1, 2, 3, 4, and multiple component systems, this "standard work on the subject" (*Nature*, London), has been completely revised and brought up to date by A. N. Campbell and N. O. Smith. Brand new material has been added on such matters as binary, tertiary liquid equilibria, solid solutions in ternary systems, quinary systems of salts and water. Completely revised to triangular coordinates in ternary systems, clarified graphic representation, solid models, etc. 9th revised edition. Author, subject indexes. 236 figures. 505 footnotes, mostly bibliographic. xii + 494pp. 5⅜ x 8.
Paperbound $2.75

A COURSE IN MATHEMATICAL ANALYSIS,
*Edouard Goursat*
Trans. by E. R. Hedrick, O. Dunkel, H. G. Bergmann. Classic study of fundamental material thoroughly treated. Extremely lucid exposition of wide range of subject matter for student with one year of calculus. Vol. 1: Derivatives and differentials, definite integrals, expansions in series, applications to geometry. 52 figures, 556pp. Paperbound $2.50. Vol. 2, Part 1: Functions of a complex variable, conformal representations, doubly periodic functions, natural boundaries, etc. 38 figures, 269pp. Paperbound $1.85. Vol. 2, Part 2: Differential equations, Cauchy-Lipschitz method, nonlinear differential equations, simultaneous equations, etc. 308pp. Paperbound $1.85. Vol. 3, Part 1: Variation of solutions, partial differential equations of the second order. 15 figures, 339pp. Paperbound $3.00. Vol. 3, Part 2: Integral equations, calculus of variations. 13 figures, 389pp. Paperbound $3.00

PLANETS, STARS AND GALAXIES,
*A. E. Fanning*
Descriptive astronomy for beginners: the solar system; neighboring galaxies; seasons; quasars; fly-by results from Mars, Venus, Moon; radio astronomy; etc. all simply explained. Revised up to 1966 by author and Prof. D. H. Menzel, former Director, Harvard College Observatory. 29 photos, 16 figures. 189pp. 5⅜ x 8½. Paperbound $1.50

GREAT IDEAS IN INFORMATION THEORY, LANGUAGE AND CYBERNETICS,
*Jagjit Singh*
Winner of Unesco's Kalinga Prize covers language, metalanguages, analog and digital computers, neural systems, work of McCulloch, Pitts, von Neumann, Turing, other important topics. No advanced mathematics needed, yet a full discussion without compromise or distortion. 118 figures. ix + 338pp. 5⅜ x 8½. Paperbound $2.00

GEOMETRIC EXERCISES IN PAPER FOLDING,
*T. Sundara Row*
Regular polygons, circles and other curves can be folded or pricked on paper, then used to demonstrate geometric propositions, work out proofs, set up well-known problems. 89 illustrations, photographs of actually folded sheets. xii + 148pp. 5⅜ x 8½. Paperbound $1.00

VISUAL ILLUSIONS, THEIR CAUSES, CHARACTERISTICS AND APPLICATIONS,
*M. Luckiesh*
The visual process, the structure of the eye, geometric, perspective illusions, influence of angles, illusions of depth and distance, color illusions, lighting effects, illusions in nature, special uses in painting, decoration, architecture, magic, camouflage. New introduction by W. H. Ittleson covers modern developments in this area. 100 illustrations. xxi + 252pp. 5⅜ x 8. Paperbound $1.50

ATOMS AND MOLECULES SIMPLY EXPLAINED,
*B. C. Saunders and R. E. D. Clark*
Introduction to chemical phenomena and their applications: cohesion, particles, crystals, tailoring big molecules, chemist as architect, with applications in radioactivity, color photography, synthetics, biochemistry, polymers, and many other important areas. Non technical. 95 figures. x + 299pp. 5⅜ x 8½. Paperbound $1.50

THE PRINCIPLES OF ELECTROCHEMISTRY,
*D. A. MacInnes*
Basic equations for almost every subfield of electrochemistry from first principles, referring at all times to the soundest and most recent theories and results; unusually useful as text or as reference. Covers coulometers and Faraday's Law, electrolytic conductance, the Debye-Hueckel method for the theoretical calculation of activity coefficients, concentration cells, standard electrode potentials, thermodynamic ionization constants, pH, potentiometric titrations, irreversible phenomena. Planck's equation, and much more. 2 indices. Appendix. 585-item bibliography. 137 figures. 94 tables. ii + 478pp. 5⅝ x 8⅜.
Paperbound $2.75

MATHEMATICS OF MODERN ENGINEERING,
*E. G. Keller and R. E. Doherty*
Written for the Advanced Course in Engineering of the General Electric Corporation, deals with the engineering use of determinants, tensors, the Heaviside operational calculus, dyadics, the calculus of variations, etc. Presents underlying principles fully, but emphasis is on the perennial engineering attack of set-up and solve. Indexes. Over 185 figures and tables. Hundreds of exercises, problems, and worked-out examples. References. Two volume set. Total of xxxiii + 623pp. 5⅜ x 8. Two volume set, paperbound $3.70

AERODYNAMIC THEORY: A GENERAL REVIEW OF PROGRESS,
*William F. Durand,* editor-in-chief
A monumental joint effort by the world's leading authorities prepared under a grant of the Guggenheim Fund for the Promotion of Aeronautics. Never equalled for breadth, depth, reliability. Contains discussions of special mathematical topics not usually taught in the engineering or technical courses. Also: an extended two-part treatise on Fluid Mechanics, discussions of aerodynamics of perfect fluids, analyses of experiments with wind tunnels, applied airfoil theory, the nonlifting system of the airplane, the air propeller, hydrodynamics of boats and floats, the aerodynamics of cooling, etc. Contributing experts include Munk, Giacomelli, Prandtl, Toussaint, Von Karman, Klemperer, among others. Unabridged republication. 6 volumes. Total of 1,012 figures, 12 plates, 2,186pp. Bibliographies. Notes. Indices. 5⅜ x 8½.
Six volume set, paperbound $13.50

FUNDAMENTALS OF HYDRO- AND AEROMECHANICS,
*L. Prandtl and O. G. Tietjens*
The well-known standard work based upon Prandtl's lectures at Goettingen. Wherever possible hydrodynamics theory is referred to practical considerations in hydraulics, with the view of unifying theory and experience. Presentation is extremely clear and though primarily physical, mathematical proofs are rigorous and use vector analysis to a considerable extent. An Engineering Society Monograph, 1934. 186 figures. Index. xvi + 270pp. 5⅜ x 8.
Paperbound $2.00

APPLIED HYDRO- AND AEROMECHANICS,
*L. Prandtl and O. G. Tietjens*
Presents for the most part methods which will be valuable to engineers. Covers flow in pipes, boundary layers, airfoil theory, entry conditions, turbulent flow in pipes, and the boundary layer, determining drag from measurements of pressure and velocity, etc. Unabridged, unaltered. An Engineering Society Monograph. 1934. Index. 226 figures, 28 photographic plates illustrating flow patterns. xvi + 311pp. 5⅜ x 8. Paperbound $2.00

## CATALOGUE OF DOVER BOOKS

APPLIED OPTICS AND OPTICAL DESIGN,
*A. E. Conrady*
With publication of vol. 2, standard work for designers of optics is now complete for first time. Only work of its kind in English; only detailed work for practical designer and self-taught. Requires, for bulk of work, no more than high school algebra and trigonometry [illegible] from fundamental concepts of geometrical, physical optics, to systematic study, design, of almost all types of optical systems. [illegible] of ordinary accuracy, accurate [illegible] [illegible] necessary [illegible] for design of [illegible] [illegible] photographic equipment. Vol. 2. Completed from author's notes by R. Kingslake, Dir. Optical Design, Eastman Kodak. [illegible] standard [illegible] [illegible] [illegible] of *Applied Optics and Optical Design* [illegible] [illegible] [illegible]. *Vol. 1*: [illegible] [illegible]. Index. Bibliography. [illegible] diagrams. [illegible] x 9¼.
Two volume set, paperbound $7.00

MECHANICS OF THE GYROSCOPE, THE DYNAMICS OF ROTATION,
*R. F. Deimel, Professor of Mechanical Engineering at Stevens Institute of Technology*
Elementary general treatment of dynamics of rotation, with special application of gyroscopic phenomena. No knowledge of vectors needed. Velocity of a moving curve, acceleration to a point, general equations of motion, gyroscopic horizon, free gyro, motion of discs, the damped gyro, [illegible] [illegible] topics. Exercises. [illegible] [illegible] [illegible]. Paperbound $1.75

STRENGTH OF MATERIALS,
*J. P. Den Hartog*
Full, clear treatment of elementary material (tension, torsion, bending, compound stresses, deflection of beams, etc.), plus much advanced material on engineering methods of great practical value: full treatment of the Mohr circle, lucid elementary discussions of the theory of the center of shear and the "Myosotis" method of calculating beam deflections, reinforced concrete, plastic deformations, photoelasticity, etc. In all sections, both general principles and concrete applications are given. Index. 186 figures (160 others in problem section). 350 problems, all with answers. List of formulas. [illegible]
Paperbound $2.00

HYDRAULIC TRANSIENTS,
*G. R. Rich*
The best text in hydraulics ever printed in English . . . by former Chief Design Engineer for T.V.A. Provides a transition from the basic differential equations of hydraulic transient theory to the arithmetic integration computation required by practicing engineers. Sections cover Water Hammer, Turbine Speed Regulation, Stability of Governing, Water-Hammer Pressures in Pump Discharge Lines, The Differential and Restricted Orifice Surge Tanks, The Differential Surge Tank, Gravity of Columns and Gates, Navigation Locks, Surges in Power Canals—Tidal Harmonics, etc. Revised and enlarged. Author's preface. Index. [illegible] [illegible] x 9¼. Paperbound $2.50

Prices subject to change without notice.

Available at your book dealer or write for free catalogue to Dept. Adsci, Dover Publications, Inc., 180 Varick St., N. Y., N. Y. 10014. Dover publishes more than 150 books each year on science, elementary and advanced mathematics, biology, music, art, literary history, social sciences and other areas.